Defining Nature's Limits

Defining Nature's Limits

THE ROMAN INQUISITION AND THE BOUNDARIES OF SCIENCE

Neil Tarrant

The University of Chicago Press CHICAGO AND LONDON

The University of Chicago Press, Chicago 60637
The University of Chicago Press, Ltd., London

Published 2022
Printed in the United States of America

31 30 29 28 27 26 25 24 23 22 1 2 3 4 5

ISBN-13: 978-0-226-81942-6 (cloth)
ISBN-13: 978-0-226-81943-3 (e-book)
DOI: https://doi.org/10.7208/chicago/9780226819433.001.0001

Library of Congress Cataloging-in-Publication Data

Names: Tarrant, Neil, author.
Title: Defining nature's limits : the Roman inquisition and the boundaries of science / Neil Tarrant.
Description: Chicago : University of Chicago Press, 2022. | Includes bibliographical references and index.
Identifiers: LCCN 2021056655 | ISBN 9780226819426 (cloth) | ISBN 9780226819433 (ebook)
Subjects: LCSH: Catholic Church—Italy—History. | Inquisition—Italy—History. | Magic—Italy—History. | Religion and science—Italy—History.
Classification: LCC BX1723 .T37 2022 | DDC 272/.20945—dc23/eng/20220104
LC record available at https://lccn.loc.gov/2021056655

♾ This paper meets the requirements of ANSI/NISO Z39.48-1992 (Permanence of Paper).

for Rachel

Contents

✷ INTRODUCTION ✷

Magic, Science, and the Counter-Reformation

In 1577, the Neapolitan nobleman Giambattista Della Porta (1535–1615) was called before the Roman Inquisition to answer allegations that he had performed necromancy. This was an ambiguous charge. While it could be taken to designate the specific act of invoking the dead to discover privileged information, it was also used more broadly to describe forms of magic that worked by making contact—whether explicit or tacit—with demons. Determining whether Della Porta had consorted with demons was made more complex by the nature of his activities. In the first book of his most famous work, *Magia naturalis* (*Natural Magic*), he argued that it was perfectly legitimate to practice natural magic. A practitioner of natural magic, he explained, was possessed of a deep philosophical knowledge of the natural order, which enabled him to manipulate and control its laws in order to create wonders beyond ordinary human comprehension. Such activities were not to be confused with harmful forms of demonic magic or necromancy, which functioned by making incantations to unclean spirits. It was the inquisitor's role to determine in this ecclesiastical court whether Della Porta's activities were in fact necromantic, and by extension whether he was guilty of the crime of heresy.[1]

The institution conducting Della Porta's trial, the Roman Inquisition, had been established in 1542. It consisted of a network of ecclesiastical courts coordinated by a central Holy Office in Rome. Initially founded to combat the spread of the Protestant heresy, the Roman Inquisition was prosecuting cases of magic with increasing frequency by the 1570s. It was concerned not simply with cases of popular magic but also with those of individuals such as Della Porta who practiced learned magical or the operative arts. These arts included activities designed to produce privileged knowledge of the future, such as geomancy, aeromancy, pyromancy, or astrology, and practical arts designed to manipulate and control the natural order, such as alchemy. Their status was disputed. Like Della Porta, many of those who practiced these arts identified as orthodox Christians and claimed that they produced their effects by natural means. Important Christian authorities had nevertheless disputed the practice of specific forms of magic since the age of the apostles, and several church fathers had condemned many of the operative arts. They had maintained that, irrespective of their operators' claims, these arts functioned only with the assistance of demons. The operators were therefore guilty of superstition—that is, the act of rendering the worship due only to God to an object or one of his creatures.[2] In this book, I explore how and why the Roman Inquisition—an institution designed to investigate charges of heresy—became involved in policing the practice of these arts.

Considering the inquisition of magic in early modern Italy requires engagement with a long and rich tradition of historical studies of the Roman Inquisition, its activities, and their consequences. From the middle of the nineteenth century—the age of the Risorgimento—Italian scholars began to formulate a distinctive interpretation of their national history. They maintained that the Italian Renaissance marked the start of the modern world and that during this period the nation's intellectuals developed habits of thought that would later give rise to the Enlightenment. For these scholars, it was essential to explain why, in the first half of the nineteenth century, Italy—the nation that had given the world modernity—remained politically divided and conspicuously unmodernized. Their answer was to blame the Church, specifically the Church of the Counter-Reformation era, which in the wake of Luther's protest had created an environment uncongenial for the exercise of free thought. While these scholars agreed on the Counter-Reformation's effects, they disputed how it caused them. Bertrando Spaventa (1817–83), for instance, argued that the Church used its organs of censorship—primarily the Roman Inquisition and Index of Forbidden Books—to suppress intellectual culture in the late sixteenth century, triggering a national decline that ended only with the advent of the Risorgimento. Others, notably Francesco de Sanctis (1817–83), argued that the Counter-Reformation in fact distracted the nation's finest minds from the task of developing the national spirit and instead encouraged them to focus their energies on protecting the institutional Church.[3]

These histories helped construct an enduring analytical framework, which I have elsewhere termed an Italian liberal historiographical tradition. Its basic concepts continued to inform the ideas and writings of eminent twentieth-century historians of varied—and sometimes changing—political persuasions, such as Benedetto Croce (1866–1952), Giovanni Gentile (1875–1944), Delio Cantimori (1904–66), and Luigi Firpo

(1915–89). These historians elaborated different aspects of the intellectual legacy that they inherited, but each sought to demonstrate the connections between the Italian Renaissance and its humanist culture on the one hand and the northern Enlightenment on the other. They maintained that the gap between these two movements was bridged by Italy's indigenous movement for religious reform—an "Italian Reformation"—and its philosophical culture. Within Italy, both movements were suppressed by the Counter-Reformation Church, with the result that Italy's finest minds, whether religious reformers or philosophers, were either silenced or forced into exile in northern Europe.[4]

These narratives of Italy's development have also informed our understanding of the history of science in early modern Italy. It has been a long-standing belief that Counter-Reformation Italy was a uniquely hostile environment in which to practice science. As I have observed elsewhere, much of the evidence adduced in favor of this proposition derives from—or at least duplicates examples used in the works of—the Italian liberal historiographical tradition. Notably, in addition to the Galileo affair, historians frequently cite as evidence for the development of an increasingly antiscientific atmosphere a cluster of trials and condemnations that occurred at the end of the sixteenth century, which included those of Giordano Bruno (1548–1600), Bernardino Telesio (1509–88), and Francesco Patrizi (1529–97). These examples also played key roles in the arguments developed by Spaventa, de Sanctis, Croce, and Firpo. It is, however, important to stress that although all of the historians of the liberal tradition subscribed to the idea of a decline of Italian culture during the seventeenth century, several considered science to be one of the few activities that allowed Italian genius to shine throughout this period of decadence.[5]

Over the last thirty years, historians have revised the older narrative of an Italian scientific decline and the Church's putative role therein. A series of studies have illustrated the continued vibrancy of Italian scientific culture in the seventeenth century, while others have shown how the Church—or at least institutions within the Church such as the Jesuit order—helped foster scientific developments.[6] Other historians have sought to revise our understanding of the Church's censorship and scrutiny of science, seeking to place notorious incidents such as the Galileo affair in a new context. The most important of these works, completed in 2009, was conducted by Ugo Baldini and Leen Spruit. Together they combed the Archives of the Congregation of the Doctrine of Faith (ACDF), identifying any documents relating to science. Their findings were revelatory. They concluded that during the cinquecento as few as three individuals were placed on trial for practicing science. The remainder of those prosecuted or censored were, they concluded, targeted either for their heterodox religious beliefs or because they were in fact practicing forms of magic or divination rather than science.[7]

In this book, I seek to engage with several of the themes that have arisen in the preceding discussion. First, I will contribute to debates about the early modern Catholic Church and its impact on Italian scientific culture. Addressing Baldini and Spruit's contention that it is possible to distinguish between magic and science in the premodern period, I argue that by censoring magic and divination the Church determined what humans could know about the natural world and placed limits on their ability to operate therein. In this respect, the Roman Inquisition exerted a greater influence on the development of science than Baldini and Spruit have allowed.[8] More broadly, I also want to consider the question of whether it is possible to correlate changes in the Church's attitudes toward ideas and knowledge-making practices relating to the natural world with the religious crisis of the sixteenth century. In short, I want to determine how, if at all, the Catholic Church's reaction to Protestantism affected the manner in which it investigated knowledge claims relating to the natural order. To answer these questions, this book places the Roman Inquisition—an institution long associated with the so-called Counter-Reformation—within a longer history. It traces the origins of this institution to the period of papal reform (c. 1000–1250), situating its development within the development of a novel program of pastoral care. It was designed to promote understanding of the faith but also had a social disciplinary aspect that involved encouraging positive behavior while discouraging errant practices. As Christine Caldwell Ames has argued, the inquisition of heresy must also be understood in this pastoral context.[9]

The Inquisition and the Counter-Reformation

The Counter-Reformation has proved to be an enduring concept. The idea of a Counter-Reformation also forms part of an influential interpretation of the development of modernity. Its key tenets were formulated by nineteenth-century German philosophers and historians who associated the advent of modernity with the Protestant Reformation while condemning the Catholic Counter-Reformation for constraining and preventing its development. Historians have challenged this interpretative framework since at least the mid-twentieth century. Notably, the German Catholic historian Hubert Jedin (1900–1980) reclaimed the term *Counter-Reformation*. He argued that while it was a defensive movement—characterized by polemics against Lutheranism and the establishment of the Roman Inquisition—it was preceded by a Catholic Reformation. He further maintained that the era of the Catholic Reformation, which began in the fifteenth century and continued into the eighteenth, bridged the medieval and modern worlds.[10] Later historians such as Wolfgang Reinhard and Heinz Schilling challenged the dichotomy between a modernizing Reformation and a retrograde Counter-Reformation by developing the concept of "confession building." They suggested that the Catholic and Protestant Churches each used similar techniques of social discipline to educate and discipline their adherents, which in turn contributed to the development of modernity.[11]

Further research has raised additional questions about the nature of the Counter-Reformation and its subsequent impact. For many historians, especially those writing about the history of early modern Italy, the Counter-Reformation has been associated with the Council of Trent (1545–63) and the creation of the Roman Inquisition (1542). Recent research has shown that Trent's significance to the subsequent development of Catholicism has been frequently misunderstood. As Simon Ditchfield has observed, historians have erroneously ascribed to Trent such fundamental changes as reform of the daily liturgy of the offices that were in fact carried out by later commissions operating under the authority of the papacy. This has, he suggested, led to the construction of "the myth of the Council of Trent as a monolithic, unified event which froze Roman Catholicism for almost four hundred years, until the thaw heralded by Vatican II (1962–65)." To understand how Trent affected the development of early modern Catholicism, it is, as John W. O'Malley has stressed, essential to identify not only what the Tridentine fathers discussed but also what they did not. Finally, as O'Malley has emphasized, Catholicism was further shaped by disputes over the interpretation and implementation of the resulting Tridentine decrees, which played out between the papacy, the episcopate, and secular rulers in the years after the council's closure.[12]

Discussion of the history of the Roman Inquisition has also been shaped by narratives of the Counter-Reformation. The Italian liberal historiographical tradition explicitly drew on the progressivist framework of nineteenth-century German historians to develop their own, similar account of their nation's history, adapting it to emphasize Italy's role in the development of modernity. Over the past forty years, much of the Italian scholarship on the early modern Church has been built around the story of the failure and defeat of the Italian Reformation. Most notably, there have been a series of important studies of the role played by the Roman Inquisition and later organs of censorship—notably, the Congregation of the Index of Forbidden Books established in 1571—in securing the Italian Reformation's defeat. Historians such as Massimo Firpo have described how

Cardinal Gianpietro Carafa, later Pope Paul IV (1555–59), used the Roman Inquisition to battle evangelical opponents within the Church.[13] Gigliola Fragnito and Vittorio Frajese have provided a series of studies of the conflicts within the hierarchy of the Church, which detail disputes between the episcopate and the Roman Inquisition about who should control censorship and determine what should be censored.[14] These studies have been complemented by Adriano Prosperi's monograph *Tribunali della coscienza*, which depicted how the Tridentine Church used instruments such as the Inquisition, confession, and missionaries to create a system of social discipline similar to that described by the theorists of confession building, albeit one that produced no beneficial modernizing side effects.[15]

While these pioneering studies have provided a compelling picture of the religious disputes of cinquecento and seicento Italy, they continue to be framed by concepts and ideas formulated within the Italian liberal historiographical tradition. Collectively, these studies provide a detailed picture of how a faction within the Church first suppressed evangelical beliefs and then constructed a novel system of social discipline. These developments established the Church's hegemony over Italian society and led to the effective suppression of any prototypical modernizing ideas or habits of thought. In these accounts, the Roman Inquisition remains a product of the *Controriforma*, conceived as a movement that emerged during the sixteenth century in reaction to the religious crisis of the age. Consequently, these accounts are focused on the events of the 1500s, with little effort made to connect the Roman Inquisition's actions to earlier inquisitorial ideas or practices, or indeed to earlier movements for religious reform. The Roman Inquisition and the Congregation of the Index of Forbidden Books therefore continue to be presented as reactions to forms of evangelism and Protestantism rather than as being inspired by any positive agenda. This reflects a wider historiographical tendency to write about either inquisition in the medieval period or the Roman Inquisition. Indeed, with the honorable exceptions of Edward Peters's *Inquisition*, Andrea Del Col's *L'Inquisizione in Italia*, and Elena Brambilla's *Alle origini del Sant' Uffizio*, historians have rarely attempted to offer a history of inquisition that bridges the two periods.[16]

In this book, I do not intend to contribute to the debate on whether it is beneficial to use the term *Counter-Reformation* to describe this period of Catholic history. On the one hand, there is insufficient room here to do justice to the richness of these exchanges or to rehearse specific ideas in any detail. On the other, by avoiding being drawn into these discussions, it is possible to develop new and perhaps more productive questions. Reformulating a proposal made by Ditchfield, instead of trying to determine what early modern Catholicism was, I will seek to reconstruct what early modern Catholics did.[17] Proceeding in this manner, I will examine the reasons why members of the Roman Inquisition prosecuted specific forms of learned magic during the sixteenth century. Adopting this approach makes it possible to consider the values and aspirations of the individuals who actually carried out the work of censorship and to reflect on how they fitted into a longer history of the Church unbound by the periodization implicit in the terminology of the Counter-Reformation or Catholic Reformation.

The activities of early modern inquisitors can be better understood if we pay attention to who they were and what they sought to achieve. In both the medieval and early modern periods, papal inquisitors were drawn primarily from the ranks of the mendicant friars and above all from the Dominican order.[18] These orders had a long history of participating in efforts to reform Christian society. The mendicants were a product of the great era of papal reform, 1000–1250. At this time, the papacy began to create a hierarchically structured Church, in which authority emanated from the pontiff and flowed through the episcopate down to the parish level.[19] From the later twelfth century, the papacy became increasingly focused on determining orthodox beliefs and practices while encouraging the development of novel techniques of pastoral care designed both to save souls and to enforce social discipline. After the Fourth Lateran Council (1215), confession and preaching assumed a new importance.[20] They allowed the Church to promote what Roberto Rusconi has termed *behavioral models* (*modelli di comportamento*). By the early thirteenth century, the newly founded orders of mendicant friars began to play crucial roles in each of these areas, frequently to the exasperation of the secular clergy. These efforts to provide pastoral care (and by extension to promote social discipline) informed the friars' activities over the following centuries, providing continuity between the medieval and early modern periods.[21] As Ute Lotz-Heumann has noted, "The most striking aspect of church discipline in post-Tridentine Catholicism is the fact that it did not so much invent new institutions and procedures of church discipline as intensify medieval ones."[22] In this book I develop this point, underlining the friars' role in transmitting to the Roman Inquisition aspirations initially formed in the context of the papal reform movement and then developed during three centuries of pastoral ministry.

This longer perspective also makes it possible to contextualize another striking feature of the early history of the Roman Inquisition. As Fragnito and Frajese have shown, the history of censorship during the later sixteenth century was marked by often bitter conflict between the episcopate and the papacy, who contested the extent of the Roman Inquisition's authority at the diocesan level. These disputes too had their roots in the era of the papal reform movement. At that time, the papacy had asserted its supremacy over the episcopate. These powers, the popes claimed, gave them the right to grant privileges to the friars that allowed them to preach, hear confession, and conduct inquisition without reference to the local bishop. From the mid-thirteenth century, secular theologians based at the University of Paris began to challenge the papacy's power to act in this manner. To do this, they contested the ecclesiological basis of the papacy's claim to supremacy over the episcopate. During the fourteenth century, these arguments provided a basis for secular rulers and members of the episcopate to challenge the papal authority. They used the arguments to question the papacy's right to delegate authority to the mendicant orders to undertake the cure of souls within parishes and, most importantly for our purposes, conduct the work of inquisition.[23] Disputes over these issues once more came to the fore during the fifteenth-century conciliar crisis, when advocates of conciliarism sought, on one hand, to reassert episcopal authority relative to that of the papacy and, on the other, to ensure that the episcopate could play a fundamental role in defining and implementing reform. These ideas subsequently informed the Fifth Lateran Council and the deliberations of Trent and later structured the sixteenth-century disputes between the episcopate and the Roman Inquisition.[24]

The Inquisition of Magic

In the early 1950s, the eminent historian Luigi Firpo (1915–89) offered an enduring interpretation of the reasons why the Roman Inquisition began to prosecute magic. His views were shaped by the Italian liberal historiographical tradition. He held that when the Church founded the Roman Inquisition in 1542 and produced its first Index of Forbidden Books in 1559, it did so to tackle a discrete problem: the spread of Protestant and philo-Protestant ideas. Having successfully completed its task by the 1590s, the Church turned this machinery of repression to examine the ideas, beliefs, and practices of Catholics. It began to examine art, literature, philosophy, science, and social mores. These investigations therefore represented an unintended, and indeed unwarranted, deviation from the Roman Inquisition's original purpose and intent. Firpo maintained that these actions suppressed the nation's intellectual culture throughout the following century, ultimately impeding Italian national development.[25] This interpretation has proved highly influential. Historians continue to argue that once the Roman Inquisition had eliminated the Protestant heresy, inquisitors began to pursue forms of deviance within Catholicism. It therefore morphed into an apparatus of social discipline that exercised control over many aspects of daily life. Echoing Firpo, some historians have suggested that this development encouraged the unjustified extension of the inquisition of heresy into areas such as literature, sexuality, and superstition.[26] Other historians have proposed the more limited thesis that the Roman Inquisition turned to the prosecution of magic only once the Protestant threat had been neutralized.[27]

In *Under the Devil's Spell,* Matteo Duni offered a distinctive variation on this latter argument. Using evidence from the Inquisition of Mantua, he suggested that inquisitors in the late fifteenth and early sixteenth centuries had regularly prosecuted cases of magic but that these investigations were halted by the sixteenth-century religious crisis. Indeed, he maintained that the centralized Holy Office's focus on Protestantism redirected the attention of local inquisitors away from the investigation of magic to such an extent that it became a "residual function of their office" and that this remained the case until 1580. He also suggested that when the prosecution of magic resumed, it did so in the changed climate of the Counter-Reformation. He noted that "following the epoch-making Council of Trent (concluded in 1563), the Catholic Church launched an unprecedented campaign to control and change the religious practices and beliefs of the faithful, bringing them into line with new guidelines and standards." For Duni, the pursuit of magic in the later sixteenth and seventeenth centuries must be interpreted within the context of a new unified post-Tridentine program of social discipline.[28]

These existing narratives about the development of the inquisition of magic are underpinned by two assumptions. The first is that in the period 1542–80 inquisitors were so preoccupied with the pursuit of Protestantism that they neglected the prosecution of magic. A lack of documentation makes it difficult to reconstruct the early activities of the Roman Inquisition in detail, but this conclusion is questionable.[29] The figures presented by William Monter and John Tedeschi in their pioneering statistical study of the Roman Inquisition show that although concern for magic increased appreciably in the later sixteenth century, it did not begin with the containment of Protestantism. Indeed, in certain regions of Italy—notably the kingdom of Naples—inquisitors appeared more concerned by magic than by Protestantism in the period 1542–1600.[30] Furthermore, as Frajese has observed, early Indices of Forbidden Books suggest that from at least the 1550s, the Roman Inquisition sought to effect a wider reformation of Catholic society by implementing such measures as controlling the circulation of the Jewish Talmud and books containing superstitious ideas and practices.[31] This evidence suggests that while controlling Protestantism and evangelical forms of Catholicism was indeed the primary focus of the Roman Inquisition's activities during the first forty years of its existence, this was not its exclusive concern.

The second assumption underpinning narratives of the development of the inquisition of magic is that the Roman Inquisition was, in effect, implementing standards of orthodoxy developed at Trent. As we saw earlier, many supposedly "Tridentine" reforms owed nothing to the council. Magic and superstition, which were not directly discussed at Trent, are a case in point. Developing Frajese's argument, I argue that rather than forming a part of a Tridentine program of social discipline, the inquisition of magic originated in an alternative program of reform conceived and implemented by the Roman Inquisition. Furthermore, I contend that these aspirations were not new but rooted in an older program of pastoral reform formulated and promoted by the mendicant orders. From as early as the thirteenth century, the friars had warned of dangers posed to Christian society by the presence of Jews and the practice of superstitious magical arts. During the fifteenth and early sixteenth centuries, mendicant preaching against both Jews and superstition intensified, as did the inquisition of magic. The surge in prosecutions of cases of superstition that occurred during the last quarter of the sixteenth century was therefore not a fundamental transformation in the purpose of inquisition. Instead, it was a shift in emphasis, which was nonetheless consistent with the actions and concerns of the early Roman Inquisition and, indeed, those of the friars.[32]

The Roman Inquisition's connections to the ideas and practices of earlier inquisitors have also been obscured by the historiography of magic and in particular by efforts to analyze the origins of the witch hunts of the sixteenth and seventeenth centuries. Examining the causes of these tragic events (and indeed the motivations underlying other historic outbreaks of violent persecution) gained a new impetus in the post-1945 period following the revelation of the horrors of the Holocaust. Historians have examined many facets of this problem, exploring underlying social and political tensions and exposing the rationale for the persecution of witches. Intellectual historians have drawn particular attention to the gradual formation of a complex of beliefs that are now referred to as the "stereotype of the witch" or the "elaborated concept of witchcraft." These histories reconstruct the evolution of an elite conception of witchcraft, tracing the formation of the idea that there existed a heretical sect of devil-worshipping witches intent on subverting Christendom. They have shown how these beliefs were developed and codified during the fifteenth century in works produced by Dominican friars, such as the *Formicarius* of Johannes Nider

(c. 1380–1438), and the notorious witch-hunting manual the *Malleus maleficarum*, written by Henricus Institoris (c. 1430–1505) and Jacobus Sprenger (c. 1436–95). Furthermore, they have demonstrated how these writings later informed the activities of witch-hunters.[33]

The tendency to privilege the development of the elaborated stereotype of the witch and the onset of the witch hunts has created two historiographical problems, one that applies to Europe as a whole and one that is specific to Italy. First, as H. C. Erik Midelfort has observed, when investigating the witch hunts, historians have tended to focus on examples of large-scale prosecutions. Although these events have been considered evidence for a so-called witch craze, they were perhaps not typical of the investigation of magic in the early modern period. Since these large-scale investigations tended to be driven by accusations of diabolic witchcraft, they have led historians to underestimate the significance of trials for "mere" sorcery.[34] Intellectual historians have introduced a similar problem. While studying texts such as the *Malleus maleficarum*, they have tended to privilege novel ideas and activities that can be assimilated into narratives of the development of the elaborated stereotype of the witch and, by extension, the witch hunts. In addition to their contributions to the development of the stereotype of the witch, the authors of the *Malleus* remained deeply concerned by "mere sorcery." As I have argued elsewhere, when they discussed these matters, they were not simply repeating established ideas but were engaged in a process of recovering and reinterpreting Thomist thought on superstition that was in and of itself intellectually innovative.[35]

Let us now consider the second historiographical problem, which is specific to Italy. Historians have tended to situate the prosecution of magic in the peninsula—and by extension the inquisition of magic—within the historiography of the European witch hunts. Analysis of the prosecution of magic in the period 1450–1600 is framed by two facts. Since at least the 1970s, historians have recognized that northern Italy was one of the first regions in Europe to experience mass hunts stimulated by the elaborated stereotype of the witch. Conversely, they have also noted that although belief in this complex of ideas persisted in the Italian peninsula, these lands did not endure witch hunts comparable to those that took place in northern Europe in the later sixteenth and seventeenth centuries.[36] Since the decline of the Italian witch hunts broadly coincided with the foundation of the Roman Inquisition, historians have suggested that these two events are causally linked. In an important contribution, John Tedeschi suggested that the Roman Inquisition prevented outbreaks of mass hunting because it enforced due process, insisted on high standards of evidence, and restrained the use of torture. While his account of the Roman

Inquisition was richly documented, Tedeschi's argument hinged on the unsupported implication that earlier inquisitorial tribunals had regularly abandoned due process. Moreover, it assumed that, had all other factors been equal, Italy was likely to have experienced witch hunts similar to those of France or Germany.[37]

In her examination of the case of the Venetian Inquisition, Ruth Martin offered an alternative explanation for the absence of witch hunts in early modern Italy. Her account stressed continuity in the history of magic. She began by tracing a history of the prosecution of sorcery and magic from the age of Augustine to the early modern period. She argued that Thomas Aquinas drew on Augustine's writings to articulate a highly influential definition of superstitious magic in his masterpiece *Summa theologica*. These ideas were subsequently incorporated into the *Directorium inquisitorum*, an inquisitorial manual produced during the fourteenth century by the Dominican friar and former inquisitor general of Aragon, Nicholas Eymerich (c. 1316–99). For Martin, the novel stereotype of the witch described in works such as the *Malleus maleficarum* represented a deviation from this established tradition that had hitherto governed the prosecution of magic. She further maintained that although there were some outbreaks of witch-hunting in Italy, the stereotype of the witch described in these texts was never widely accepted. Indeed, by the early sixteenth century, "ecclesiastical pronouncements on witchcraft return[ed] more to their original mould." When the reorganized Venetian Inquisition began its work, its inquisitors' activities were guided by Eymerich's manual, the *Directorium inquisitorum*, which contained a conception of magic that pre-dated the innovations contained in texts such as the *Malleus maleficarum*.[38]

Although Martin may have underestimated the diffusion of the stereotype of the witch in late fifteenth- and early sixteenth-century Italy, her work offered a valuable means to interpret the connections between the practices of medieval inquisitors and those who worked in the Roman Inquisition. Her central point was echoed by Francesco Beretta, who in a seminal article suggested that to understand censorship in the seventeenth century, it is necessary to reconstruct the criteria used by the Roman Inquisition. Beretta's article identified the concepts that guided the Inquisition's oversight of philosophy, tracing them principally to Thomas Aquinas's synthesis of knowledge.[39] Expanding Martin and Beretta's approach, in this book I reconstruct the principles that governed how members of the Roman Inquisition conceptualized, understood, and justified the prosecution of magic. Like Martin and Beretta, I argue that Aquinas's ideas were fundamental. His discussion of superstition, contained primarily in the *Summa theologica*, informed the investigations of medieval inquisitors and continued to provide the intellectual basis for the Roman Inquisition's investigations of magic, including its learned forms.

The reception of Aquinas's ideas was, however, more complex than this bald summary would suggest. It is difficult to maintain that there was an essentially consistent "Thomist" tradition of thought on superstition. Aquinas wrote widely on the problem of superstition over the course of a productive career, often changing and refining his opinions on the validity of specific operative arts. Building on a number of my previous articles, I argue that Aquinas's writings could be used to construct and validate a range of sometimes contradictory "Thomisms." In the *Directorium*, for example, Eymerich elaborated a recognizably Thomist concept of superstition, but it was not one directly informed by the *Summa theologica*. By contrast, Dominicans more closely associated with the elaborated stereotype of witchcraft—Nider, Institoris, and Sprenger—were at the forefront of a fifteenth-century revival of Thomist thought. Their writings on superstition were directly informed by a reengagement with the concepts contained in *Summa theologica*, and they encouraged contemporary inquisitors to investigate forms of superstition that Eymerich had previously ignored. When the Roman Inquisition was established in 1542, there existed at least two related yet distinct sets of "Thomist" criteria pertaining to superstition. By drawing on each of these contradictory sets of criteria, members of the Roman Inquisition created a series of ambiguities in the censorship of magic that would persist into the seventeenth century.

By reconstructing these criteria, it is also possible to make a significant contribution to debates about the relationship between magic and modern science and how these boundaries were negotiated during the seventeenth and eighteenth centuries. Such questions once again have clear roots in discourses about modernity. In Max Weber's celebrated formulation, modern society gradually became "disenchanted" as erroneous and misplaced belief gave way to reason and science.[40] Historians continue to debate the relationship between science and magic, whether they were connected, and, if so, why and when they separated. As we have seen, Baldini and Spruit sought to distinguish between modern science and magic, rendering the Roman Inquisition's censorship of these arts irrelevant to the development of modern scientific disciplines. Others have offered the basis for more nuanced analysis. In an important paper, John Henry argued that during the seventeenth and eighteenth centuries some parts of magic (for example, the alchemists' ambition to manipulate and control materials) became incorporated into natural philosophy and ultimately modern science, while a rump of activities (for example, the transmutation of base metals into gold) came to be considered spurious and associated with our modern idea of magic. He further argued that the various churches hastened this process, because they opposed the practice of all forms of magic.[41]

While useful, Henry's account needs to be refined. First, the Catholic Church's position on magic was more complex than mere opposition. Furthermore, there remains a danger of interpreting the story of the inquisition of magic teleologically—that is, to see Catholic censorship fostering the development of modern science by hastening the process of disenchantment. While the Catholic Church certainly sought to use its centralized organs of censorship to eliminate superstitious practices, these actions reiterated inquisitors' belief in a cosmology in which magic represented a real and viable hazard. Indeed, they considered the suppression of superstition to be important precisely because demons existed and presented an ongoing danger to Christian souls. Moreover, as I shall argue in this book, although the Roman Inquisition considered many specific arts superstitious, it was equally opposed to any philosophical accounts of nature that denied the possibility of magic. The truth of magic was underwritten by orthodox Christian belief in the existence of angels and demons, and any attempts to deny the existence of magic tout court would present a profound challenge to these tenets of the faith. For these reasons, the Church wished to suppress the practice of certain magical arts while at the same time demanding that Christians acknowledge both the reality of their existence and the danger that they posed.

Rather than seeking to remove magic from Catholic cosmology, the members of the Roman Inquisition were engaged in a process of determining the boundaries between orthodox and superstitious ideas and practices. Consequently, they did not seek to suppress all of the operative arts. Like many of their fellow Christians, they accepted that some of them were perfectly legitimate. Inquisitors were therefore forced to determine which arts should be permitted and the basis for making these decisions. Key to many of the debates explored in this book was the question of whether a particular art—for instance, alchemy, astrology, physiognomy, or chiromancy (palm reading)—could naturally produce the effects that its operator claimed. Drawing on the approaches developed in the work of scholars of demonology such as Stuart Clark and Robert Bartlett and historians of science such as William R. Newman, I explore how philosophical knowledge of the natural world was enrolled in efforts to determine the orthodoxy of specific operative arts.[42] To form such a judgment, inquisitors were obliged to pronounce on the validity of the knowledge claims that justified a particular art's practice. They had to determine whether the operator's understanding of the manner in which the natural world functioned was or could be true. Since it was invested with the authority to pronounce on these issues, the Roman Inquisition's primary effect on the history of science was to determine the possible explanations of the natural order available to Italian intellectuals.

Defining Nature's Limits

The purpose of this book is to examine how the Roman Catholic Church set the parameters for making knowledge about the operations of the natural order in the period 1000–1564. It is in this sense a contribution to the history of science, although it does not always concern material that some modern scholars would consider "scientific." The society that we are considering in this book held assumptions about the natural order that were radically different from those presupposed by modern science. Premodern Christians sought to understand a cosmos that they knew was created by God, which they believed that he had populated not only with humans and animals but also with angels, some of whom had fallen to become demons. Before about 1750, and perhaps later, intellectuals who wished to offer a true account of the natural order often felt compelled—or were obliged—to take account of the actions of God and his creatures such as angels and demons. The potential explanatory repertoire that premodern intellectuals could draw on to comprehend observed or reported phenomena within the natural order was therefore far wider than that allowed by modern science.

That premodern Christian intellectuals accepted this wider interpretative field did not mean that their ideas were irrational or that they always understood the natural order in terms of divine, angelic, or demonic causation. At the heart of most philosophical accounts of nature in the premodern era lay the idea of the cosmos, a depiction of the natural world derived from Greco-Roman philosophy. The majority of scholars believed that God's creation followed fixed, predictable rules, which humans could comprehend, albeit imperfectly. The tenets of their faith nevertheless required Christians to acknowledge the existence of a providential God and his creatures, including angels and demons. The acceptance of these beliefs in turn compelled them to accept that the rules that ordinarily governed the operations of the natural order could be disrupted. God, and God alone, could break the rules that he had established in order to perform miracles—that is, to produce naturally impossible effects such as restoring the dead to life. Since angels and demons were God's creatures and hence part of the natural order, they could not break his laws. They nevertheless understood and possessed a capacity to manipulate God's laws that surpassed that of humans. Consequently, they could produce wonders that defied human understanding. To a lesser extent, humans too could manipulate these laws to produce their own wonders. Humans could also solicit the assistance of angels or demons to perform works on their behalf. During the medieval and early modern periods, Christian intellectuals—including the inquisitors discussed in this book—sought to determine whether any particular effect or particular art that purported to produce that effect could be attributed to natural causes (the routine functioning of nature's laws or their legitimate manipulation by humans), preternatural causation (the manipulation of nature's laws by angels or demons), or supernatural causation (God's direct action to break the laws of his creation).[43]

As Stuart Clark has noted, premoderns' willingness to attribute effects to either preternatural or supernatural causation did not represent an acknowledgment of their failure to comprehend those effects. On the contrary, for contemporaries, to be able to assign a given event to either preternatural or supernatural causation was to know its causes. The central problem facing Christians in both the medieval and early modern periods was determining to which category of causation a given event or effect should be attributed. They needed to know whether a particular event or effect could be produced by the ordinary working of nature, whether it had been naturally induced through human artifice, or whether it lay beyond human capabilities. If the event or effect was considered to be beyond ordinary human capacities, it raised the possibility that the person who had created that effect had solicited the assistance of demons and was therefore liable to censure by the Church. Consequently, it was essential to understand precisely what effects could be produced by unaided humans and which effects were so remarkable that they could be produced only with the assistance of demons. To make these determinations, contemporaries had to demarcate the limits of human ability to understand and operate in nature, a task that they accomplished by drawing on a range of resources including not only theology but also natural philosophy.[44]

Studying the inquisition of magic offers a means to explore how medieval and early modern Christians understood and defined nature's limits and the means by which they disseminated and enforced them. The investigation of learned magic forced inquisitors to situate the effects that an operator claimed to produce on an explanatory grid that included natural, preternatural, and supernatural causation. The boundaries between these categories were often unclear, but more importantly the criteria used to make these judgments were contested. In the medieval and early modern periods, contemporaries practiced, theorized, criticized, and feared a wide range of different magical arts. These included forms of magic—both learned and popular—that explicitly required making contact with demons, those that intentionally subverted Christian ritual, and those that involved groups of erstwhile Christians gathering to worship the devil. While this book discusses numerous forms of magic in the premodern period, it focuses on a smaller set of learned operative arts. The operators of such arts as making astrological talismans or chiromancy claimed that although they were able to perform wonders beyond ordinary comprehension—whether manipulating nature to produce extraordinary effects or to procure privileged knowledge of the future—they made no explicit appeal to angels or demons and their arts included no ritual element. They claimed instead that they worked their wonders naturally. Such arguments were rejected by some within the Church on the grounds that they could not work by natural means. By seeking to regulate the practice of these disputed arts—a subsection of the field of magic—the Church defined the legitimate boundaries of natural knowledge. To put it anachronistically, they defined the limits of science.

My central claim in this book is that before the establishment of the Roman Inquisition in 1542, the criteria used to determine the boundaries between legitimate human activity and illicit preternatural magic were subject to debate within the Catholic Church. Before this time, there was no universal standard for judging the legitimacy of ostensibly natural arts. There were, however, clear arguments rejecting them. I argue that the intellectual and institutional roots of the Roman Inquisition's prosecution of magic, and especially the prosecution of ostensibly natural arts, can be traced to the writings of Thomas Aquinas. After 1542, the Roman Inquisition began to impose a broadly coherent set of criteria derived from the thought of Aquinas that was—in principle at least—binding on all Catholics. In this manner, they made important strides toward defining nature's limits. The incorporation of Aquinas's ideas into inquisitorial practice was, however, complex. For nearly 250 years, Aquinas's discussions of the legitimacy of arts such as astrological talismans and chiromancy were, at best, marginal to inquisitorial practice. In this book, I trace how these Thomist principles became established as the criteria of censorship in the Roman Inquisition and, by extension, how they subsequently became regarded as the standard of orthodoxy in the Catholic Church.

Ideas are of central importance to this story. Yet as historians of science have stressed for over fifty years, explaining why members of a particular society regarded ideas to be authoritative (irrespective of whether they are now considered "true" or "false") requires us to examine factors seemingly external to the ideas themselves. Ideas, they have stressed, are produced by humans working within specific social, cultural, and political contexts, who are driven to act by an array of imperatives including but not restricted to disciplinary formation, institutional expectations, intellectual curiosity, personal ambition, and religious fervor. The historian's task is to reconstruct the reasons why past actors sought to legitimate and gain universal assent for ideas produced in often highly localized contexts and the strategies and techniques they deployed in pursuit of these ends.[45] Adopting a similar approach, this book presents two interrelated stories. The first is essentially an intellectual history of Thomist thought on magic, which traces changing interpretations of Aquinas's ideas within the Dominican order between the thirteenth and sixteenth centuries. The second is an account of the inquisition of magic. It describes how members of the Dominican order first incorporated a Thomist conception of magic into the practice of medieval inquisitors, thereby forming an investigative framework that later Dominicans would incorporate into the censorial structures of the sixteenth-century Church. Tracing this latter story requires us to situate Dominicans' involvement in the inquisition of magic, charting it against a long history of ecclesiastical conflict and reform that dates to at least the age of Gregorian Reform.

This book is divided into two parts. The first treats the establishment of the intellectual and institutional foundations on which the Roman Inquisition was later built. It locates their origins in the period of the papal reform movement (c. 1000–c. 1250) and traces their development to the eve of the Great Schism of 1378. At the heart of this story lies an account of the process by which the papacy exerted its power to define orthodox belief and to enforce those standards over Christians living in western Europe. During this period, the papacy placed itself at the forefront of a reform program that explicitly sought to create a homogeneous Christian society purged of deviant thought and errant behavior. This process entailed both the elaboration of a set of orthodox ideas and behaviors and the construction of a system of pastoral care to educate and discipline the people of Christendom. To help achieve these ends, the papacy oversaw the creation of two novel institutions: the orders of mendicant friars and the office of the inquisitor of heretical depravity. This section of the book describes how papal inquisitors acquired the authority to investigate cases of magic, before surveying subsequent debates over inquisitors' authority to intervene in these cases and discussions of the criteria that should be used to determine the boundaries of orthodoxy.[46]

Chapter 1 examines the origins of the inquisition of magic. This was a novel practice, which depended on two key developments. First, it required the creation of a cadre of papally appointed officials invested with the authority to investigate allegations of heresy. The papacy acquired the power to intervene within the dioceses of other bishops as a result of the reform movement of the eleventh and twelfth centuries. During these years, the papacy asserted its supremacy over the western Church. It claimed the right to lead the reform of Christian society by defining orthodox thought and behavior and to have the authority to appoint officials to intervene in individual dioceses irrespective of the wishes of local bishops. The primary means for these interventions were the actions of the mendicant friars, who delivered pastoral care and performed the role of inquisitors of heretical depravity. Second, the inquisition of magic depended on the expansion of the definition of the term *heresy*. Heresy was in essence a thought crime. It entailed the conscious decision to maintain a belief contrary to the faith. Before the fourteenth century, magic was conceived not as a heresy but as a superstition—that is, an act that offered undue worship to something other than God or the saints. Church fathers such as Augustine and Isidore of Seville had prohibited the practice of virtually all operative arts on this basis, a view that was reiterated in canon law during the eleventh and twelfth centuries. To facilitate the inquisition of magic, the papacy ruled in 1258 that certain forms of magic could be prosecuted as heresy. Both of these developments were contested and, as we shall see, formed the basis of disputes about the inquisition of heresy until at least the sixteenth century.

The papacy's decision to grant inquisitors of heretical depravity the authority to investigate certain categories of magic was just one in a series of transformations in the way magic was understood in the high medieval period. Chapter 2 describes how these changes were driven by the rapid intellectual developments of the twelfth and thirteenth centuries, during which theologians and philosophers rediscovered not only the intellectual legacy of pagan antiquity but also the church fathers' strategies for dealing with that legacy. The mendicant orders played a central role in this process by developing novel ways to employ philosophy in the service of the faith. Two Dominican theologians, Albertus Magnus and Thomas Aquinas, offered innovative means to resolve conflicts between knowledge claims established in the disciplines of philosophy and theology. They also wrote extensively on the issue of magic, developing a novel rationale for identifying legitimate operative arts that was rooted in a philosophical analysis of the natural order. They contended that it was legitimate for Christians to practice an art if it functioned by demonstrably natural means. In texts such as the *Summa theologica* and the *Summa contra gentiles*, Aquinas articulated a clear rationale to defend the practice of a range of arts (including predictive astrology) that had been rejected not only by the church fathers but also by the canonists of the eleventh and twelfth centuries.

By the beginning of the fourteenth century, the papacy had asserted its right to appoint inquisitors to investigate cases of heresy, including heretical magic, across Christian Europe. These new powers were not universally accepted, however. Chapter 3 describes how secular and episcopal authorities contested the papacy's authority not only to appoint inquisitors in local dioceses but also to allow the mendicants to conduct the cure of souls. It also shows how the papacy responded to these challenges by reasserting the inquisitor's authority to investigate heresy and by formulating a novel definition of heretical magic. In the bull *Super illius specula*, the papacy argued that anyone who performed arts that involved making an explicit compact with demons should be treated in the same manner as heretics. The papacy subsequently encouraged the prosecution of magic in all courts whether episcopal, civic, or inquisitorial, but it met with limited success. The desire to assert papal authority and further the prosecution of illicit operative arts informed Nicholas Eymerich's discussion of magic and divination. In his *Directorium inquisitorum*, he drew on the papal bull and an idiosyncratic reading of Thomas Aquinas's work to assert the inquisitors' right to investigate cases of magic. The criteria he used to define the boundaries of magic owed little to the ideas that Aquinas had developed in the *Summa theologica* or the *Summa contra gentiles*. Eymerich instead urged the inquisitor to establish whether the operator had explicitly invoked demons to work their effect—that is, to investigate their actions rather than to consider whether the art in question was naturally possible. Consequently, Aquinas's efforts to identify illicit magic by defining nature's limits remained marginal to inquisitorial practice. In any case, despite the papacy's efforts, the prosecution of magic remained rare in the courts of late fourteenth-century Europe.

The second part of the book focuses on continuity and change in the Dominican order's intellectual conception of magic and on the intensification of its efforts to combat superstition between the fifteenth and sixteenth centuries. It places these developments within the context of the radical challenges posed to papal authority by the emergence of conciliar theory in the fifteenth century, which questioned not only the papacy's right and ability to guide reform but, in some formulations, also the basis of the papal supremacy. Perhaps the most lasting legacy of conciliarism was the development of an alternative model of reform centered on a reinvigorated episcopate dedicated to the delivery of pastoral care. To rebuild its authority within the Church, the papacy sought to direct and promote an alternative program of reform. Successive popes encouraged the development of the Observant reform movement that had begun in the later fourteenth century, which sought to transform not only the religious orders but also Christian society.[47] In support of the latter aim, the popes encouraged the Franciscan and Dominican orders in their efforts to build on their earlier social disciplinary techniques to launch their own program of reform during the fifteenth and sixteenth centuries. This program sought to encourage devotion among ordinary Christians while protecting them from the threats posed by heresy, the presence of the Jews, and superstition. These ends were to be achieved by using means of pastoral engagement such as preaching, hearing confession, and, where necessary, the inquisition of heresy. The religious crisis of the mid-sixteenth century brought to the fore tensions between the rival episcopal and papal models of reform. As we shall see, they were also clearly evident during the Council of Trent. In addition to defining Catholic doctrine, the Tridentine fathers debated the nature of reform, how and by whom it should be organized, and the precise ends to which it should be directed.

From the later fourteenth century, papal authority was undermined by two related crises: the Great Schism and the rise of conciliarism. Chapter 4 describes how the supporters of conciliarism mobilized long-standing complaints to curtail papal power and to bolster episcopal authority by articulating alternative proposals for Church governance and reform centered on the figure of the bishop. After the resolution of the crisis in the mid-fifteenth century, the papacy used a number of means to rebuild its authority. These included encouraging the spread of the Observant movement, which in turn allowed the mendicants to enhance their pastoral mission. Expanding their ministry, the friars intensified their programs of pastoral engagement, preaching with increased vigor against the threat posed by Jews and Judaizers, heresy, and superstition. The friars' renewed interest in superstition was paralleled by radical intellectual changes within the Dominican order. On the one hand, authors such as Johannes Nider began to articulate a new idea of diabolic witchcraft, beliefs that inspired a new wave of trials for magic in Switzerland, France, and Italy. On the other, theologians within the order also engaged in a systematic fashion with Thomas Aquinas's ideas on superstition, in the process recovering his central contention that understanding nature's limits was central to any attempt to define the boundaries of superstition. These ideas were incorporated into inquisitorial manuals such as the *Malleus maleficarum* for the first time in the late fifteenth century. Initially, these developments had only a limited effect on the inquisition of magic. Trials continued to focus on allegations of diabolic witchcraft and forms of superstition that would have been recognizable to Nicholas Eymerich rather than the practice of ostensibly natural arts.

Chapter 5 takes up the story from the beginning of the sixteenth century. By this time, tensions between rival proposals for reform were becoming increasingly evident. The mendicants continued to pursue their pastoral ministry throughout Europe. Within Italy, their campaign against diabolic witchcraft reached a new intensity in the 1520s. Although a significant number of these trials involved allegations of diabolic witchcraft, many inquisitors also continued to prosecute older forms of superstition. During the same period, some Catholics proposed reforms to doctrine, while others also suggested reforms to the Church in head and members. Many of these proposals centered on the episcopate, and in some instances they also involved the curtailment of mendicant privileges. These proposals held significant implications for the authority of the papacy relative to that of the episcopate. Disputes over these issues were played out at the Fifth Lateran Council (1512–17). To ensure that the meeting could conclude successfully without undermining papal authority, Pope Leo X carefully balanced the contrasting demands of the mendicants and the episcopate. The council fathers passed a series of reform decrees, many of which related to the office of the bishop and provided a basis for later Tridentine reforms. Although several of the Lateran decrees reiterated the importance of the mendicants' ministry generally and encouraged the pursuit of magic specifically, they did not offer any changes to current conceptions of magic. The inquisition of magic continued to proceed in the same manner as it had at the beginning of the century, focusing on the prosecution of diabolic witchcraft and sorcery.

In 1517, Christian Europe was divided by the actions of the Augustinian friar Martin Luther. All too often, the Catholic response to these events has been reduced to a model of definition and enforcement, which posits that Catholic doctrine and practice were established at the Council of Trent and imposed by the Roman Inquisition. As we have seen, however, many of the ideas and policies taken to be characteristic of post-Tridentine Catholicism did not originate in the council. Chapter 6 shows that between the 1520s and 1540s there emerged two distinct approaches to addressing both the Lutheran challenge and the reformation of the Church. The first, a conciliar approach, was supported by the emperor and a number of bishops with dioceses in his territories. It led to the convocation of the Council of Trent, which the Lutherans were invited to attend, and where fathers discussed doctrine and set out a program of bishop-led reform. Proponents of the second approach to the Lutheran crisis rejected the idea of conciliarist reform. High-ranking clerics such as Cardinal Carafa denied the need for a council to effect institutional reform, refused to countenance doctrinal reform, and objected to the idea of negotiating with Lutherans. They instead proposed that the papacy should direct a limited series of institutional reforms while rejecting Lutheran doctrine and rooting out all who held these or similar ideas. Pursuing these ambitions, Carafa convinced Pope Paul III to allow him to reorganize the inquisition of heresy. This led to the creation of the Roman Inquisition, a development that gave Carafa immense power but also strengthened the mendicants' influence and authority within the Church, especially that of the Dominicans. As we shall see, members of this order used the Roman Inquisition to pursue their own program of reform, which sought to deal with issues that were never considered at Trent, such as the problem of superstitious magic.

By the beginning of the 1550s, the first period of Trent had been dissolved rather than formally concluded. The ecclesiastical hierarchy remained divided by two differing visions of reform, one papal and the other conciliar. While the council again met inconclusively between 1551 and 1552, Carafa continued to use the Roman Inquisition throughout the early 1550s to protect his vision of Catholicism—that is, one led by the papacy and purged of heresy. In 1554, he ascended to the papal throne as Paul IV, quashing any hope that the council would be reconvened and strengthening the power of the Roman Inquisition. It is well known that during these years, Paul IV used the Inquisition to define Catholic doctrine. Chapter 7 shows, however, that the Dominican friars also used the Roman Inquisition's new power and authority to advance their own reforming agenda. Using instruments such as the Index of Forbidden Books, the Dominicans sought to impose their attitudes toward the Jews, philosophy, and magic on Christian society. Analysis of the Inquisition's indices also reveals that the Roman censors used the criteria first developed by Aquinas and subsequently rediscovered in the fifteenth century to define the boundaries of orthodoxy. Upon Paul IV's death, supporters of the conciliar approach reasserted their authority, leading to the reconvocation of Trent. During this meeting, the fathers of Trent sought to roll back the excesses of Paul IV's regime and once more assert episcopal control over reform. These developments moderated the Inquisition's power, but the fathers endorsed a new index that once again accepted Aquinas's criteria for censoring magic. There was now a broad consensus that in order to determine the legitimate boundaries of superstition, it was necessary to determine nature's limits.

I

Medieval Foundations

✷ 1 ✷

The Origins of the Inquisition of Magic

Introduction

In 1258, Pope Alexander IV (1254–61) issued the bull *Quod super nonnullis*, which declared that inquisitors held the power to investigate forms of magic that "manifestly savored" (*saperent manifeste*) of heresy.[1] Its promulgation was a turning point in both the history of the Inquisition and the history of magic. From a jurisdictional perspective, the bull expanded the inquisitor's competence. It invested appointed officials with the authority, if not the duty, to investigate allegations of heterodox magical practice. From an intellectual and legal perspective, Alexander IV's judgment was a crucial stage in the process of constructing magic as a form of heresy. Before this time, ecclesiastical authorities had condemned specific magical practices as superstitions, which was a category distinct from heresy. The fact that Alexander could invest inquisitors with the power to investigate magic reflected a radical transformation in the papacy's authority. From the mid-eleventh century, a series of reforming popes began to assert the papacy's supremacy, its right to define orthodoxy within the Church, and its power to reorganize pastoral care at the diocesan and parochial levels. By the beginning of the thirteenth century, the papacy had also created the office of Inquisitor of Heretical Depravity, which enabled it to send appointed officials directly to investigate heresy in specific regions. None of these developments were inevitable. They were the product of a series of interrelated developments that took place between the late tenth and the mid-thirteenth centuries—that is, during Latin Christendom's age of reform.

The Intellectual Revival of the Latin West

Christians' capacity to think about, practice, and contest magic and the operative arts was framed by the intellectual resources that they had to hand. During the patristic era, knowledge of various operative arts and the philosophical ideas that supported them remained widely diffused throughout the empire. The availability of these resources prompted fathers such as Augustine to write about these arts and facilitated their critiques. After the collapse of the Western Roman Empire in the late fifth century, this knowledge was gradually lost. It would be highly misleading to talk of an absolute intellectual decline during this period, however. The intellectual life of the successor states remained vibrant and was revitalized during the Carolingian era (c. 750–c. 875) by the creation of monastery and cathedral schools. Scholarship nevertheless tended to focus on scriptural studies; consequently, secular learning and especially knowledge of Greek philosophical accounts of the natural world diminished.[2] Scholarship was further disrupted by the invasions of the tenth century and the political, social, and economic upheavals that followed in their wake.[3]

These disruptions were relatively short-lived. By the eleventh century, monastery and cathedral schools founded in the Carolingian period began to flourish once more, while institutions such as the medical school at Salerno, established in the ninth century, began to enjoy a wider reputation. There were also innovations, notably the foundation of the universities.[4] There were also intellectual changes. Like the later Renaissance, the scholarship of this period was stimulated by its protagonists' desire to restore the learning of the ancient world. By rediscovering texts that contained Greco-Roman scholarship—and, in this instance, a wealth of Arabic learning—contemporaries were able to produce major innovations in such fields as literature, law, theology, medicine, and philosophy.[5] This engagement with ancient and contemporary Arabic-language philosophy also exposed medieval Christians to sophisticated accounts of the cosmos, which in turn fostered innovative ways of thinking about the natural world. Many historians consider these developments to be major contributions to the development of science.[6]

Although Christians made a conscious effort to restore the learning of the Greco-Roman world, the disciplines of knowledge and the learned practices that they created differed from those of antiquity.[7] Since there were few channels for the direct transmission of Greco-Roman scholarship, twelfth-century scholars reconstructed ideas and practices from texts that had been rediscovered or translated from Arabic copies. Aristotle's works, for example, had been virtually unknown in the Latin West, but by circa 1200 most of his extant corpus had been translated into Latin. To learn how to comprehend these complex works, Christians frequently turned to Arabic commentaries. Their understanding of these texts was therefore mediated not only by the process of translation but also by the interpretations offered in the commentaries and original works of Arabic-language scholars.[8] From these varied sources, medieval Christians began to create new disciplines of knowledge and practical arts such as physick. As both a body of knowledge and a practical activity, physick was derived from ancient ideas and practices, most obviously the form of natural explanations and therapeutics developed within the Hippocratic tradition. Yet medieval Christians, influenced by Arabic-language authors such as Avicenna, developed a particular form of humoralism rooted in Galen's later systematization of Hippocratic thought. Physick was therefore a new form of humoral healing that was based on yet distinct from both its textual sources and ancient healing practices. Christians repeated this process of appropriation and use in other areas of endeavor, including philosophy, theology, and law.[9]

As a result of the intellectual ferment of the eleventh and twelfth centuries, some Christians began to engage with pagan and Arabic-language texts that described ideas and practices such as astrology, alchemy, and theurgy. Knowledge of these and other operative arts was not unprecedented. A variety of "magical" ideas and practices had been common at all levels of society in early medieval Europe, and some continued to be influenced by ancient culture.[10] For example, throughout the early medieval period, Christians continued to practice astrology, an art rooted in ancient ideas and practice. Nevertheless, their understanding of the theoretical basis of astrology tended to be relatively limited. The intellectual developments of the twelfth century allowed Christians to expand and enrich their understanding of this art and develop their technical capability to make predictions. Astrology therefore exemplifies how Christians either acquired or improved their capacity to use ancient operative arts by engaging with ancient and Arabic textual traditions. No less than the development of learned medicine or the use of formal logic, the construction of Christian forms of learned magic was the result of a self-conscious attempt to rediscover, emulate, and repurpose the learning of the Greco-Roman world. In this respect, medieval Christians' recovery of learned magic reflected their increasing intellectual sophistication and ambition every bit as much as their newfound ability to offer an Aristotelian account of motion.[11]

Broadly speaking, Christians engaged with arts drawn from two main categories of learned magic. For brevity, we can refer to the first with the etic term *natural magic*. The practitioners of these arts claimed that they could produce privileged knowledge or extraordinary effects because they possessed a sophisticated philosophical understanding of nature, which allowed them to predict, manipulate, or control the laws of the cosmos. Once again, astrology provides a good example. Its proponents believed that through the study of the movements of the stars they could produce privileged knowledge of the future. While some contended that it could be used to predict future human actions, others believed it made knowledge useful for vital activities such as agriculture, navigation, and, above all, medicine. Indeed, within the universities, astrology developed in tandem with medicine, as scholars began to explore its predictive capacities and, in some instances, to consider the potential uses of astronomical images.[12] Other forms of natural magic rediscovered in this period included geomancy, pyromancy, and physiognomy.[13] Christian scholars also began to encounter the Arabic tradition of alchemy, a practical art intended to fulfill a range of purposes including chrysopoeia (the art of transmuting metals) and the production of medicaments and distillations.[14]

During this period, Christian scholars also became increasingly interested in another broad category of arts: preternatural magic. These were theories and techniques that explicitly depended on the invocation of either angels or demons. Since these arts required advanced education, they were often practiced by priests who constituted what Richard Kieckhefer has referred to as a "clerical underworld."[15] The practices recorded in *Ars notoria* are perhaps the most celebrated—or infamous—example. This text probably dates to the late twelfth century and most likely originated in Bologna. Although the art described in this text resembled forms of theurgy practiced in Late Antiquity, they were not, strictly speaking, the same. The *Ars notoria* involved the use of elaborate cycles of prayers to beseech angels for their assistance. Specifically, and no doubt reflecting the fact that it was produced in a university town, they were designed to enable students to acquire knowledge of the liberal arts rapidly and inexpensively. Many Christians condemned this art, believing that it constituted a form of illicit necromancy—that is, an act that involved deliberately and explicitly contacting demons to achieve specific ends. For its practitioners, however, the techniques described in *Ars notoria* were fully compatible with Christian orthodoxy.[16] By no means were all theurgic practices designed to contact angels. Other Christians did explicitly seek to control demons. Some defended this practice by claiming that, like King Solomon, they could command these evil spirits through God's power. Others drew comparisons with exorcism, a rite that enabled Christians, with God's assistance, to compel a demon to leave a body that it possessed. Certain necromantic practices did explicitly involve offering sacrifices to demons, however.[17]

Magic and the Law

The rapid development of interest in the operative arts forced medieval Christians to consider not only which arts, if any, they could legitimately practice but also the correct basis for making these judgments. Medieval Christian society did not, however, possess a single universally recognized authority, whether textual or institutional, to guide these decisions. After the disintegration of the Western Roman Empire in the late fifth century, authority in the Latin West had fragmented; government varied by region, and authority was shared between secular and religious powers. Although secular and religious authority were closely interrelated, secular law began to develop separately from ecclesiastical law, with important implications for the prosecution of magic. Secular authorities tended to be concerned by forms of magic that produced harmful effects—that is, maleficiant magic. The punishments they imposed for engaging in such practices could be as severe as execution. They were, however, relatively unconcerned by many common forms of magic, such as love magic or spells designed to locate lost property, which were not overtly maleficiant. They also paid little attention to ostensibly benign operative arts that purported to produce wondrous effects.[18]

Ecclesiastical law developed differently. Before circa 1000, the Church possessed a series of rules pertaining to doctrine, discipline, and ecclesiastical structures but lacked a single coherent body of law that could be readily consulted. Existing Church law was drawn from a variety of sources. Scripture, which condemned magic in numerous passages, provided its foundations. These texts were supplemented by the writings of church fathers such as Augustine (354–430) and Isidore of Seville (560–636). These fathers not only condemned numerous forms of magic but, more importantly, established the principle that when considering the legitimacy of a particular art, its intended purpose—whether beneficent of maleficiant—was unimportant; what mattered was how it was achieved. Augustine, especially, argued that to produce any knowledge, arts such as predictive astrology, hydromancy, and pyromancy required their operator to form a pact with demons. These acts were therefore superstitious, in the specific sense that they required the operator to offer the reverence due only to God or the saints to another of God's creatures. Ecclesiastical law was also informed by a body of judgments issued by bishops, legislation produced by local church councils or synods, and judgments taken by the universal or ecumenical councils. These various sources provided a wealth of material, but it was dispersed and sometimes contradictory. Although there were some attempts to compile unofficial collections of canons, it remained difficult to consult the existing body of laws on specific matters, including the legitimacy of magic.[19]

Attempts to systematize canon law began during the eleventh century. At this time, individuals such as Burchard of Worms (c. 950/65–1025) collated canons produced by provincial or universal councils and the writings of the church fathers so that bishops could refer to them when producing judicial decisions on specific subjects. Burchard's work, *Decretum*, was composed of twenty thematically arranged books. Book 10, "De incantatoribus et auguribus," dealt with the problem of magic and divination. The opening canons, drawn primarily from conciliar legislation, tended to concern the correct punishment for magical offenses. The first text Burchard cited was the canon *Episcopi*, a ninth-century document that he incorrectly ascribed to the Council of Ancyra (314). It described the activities of the "Riders of Diana," a group of individuals who believed they had been transformed into beasts, although the canon suggested that they were the victims of demonic illusion. This text would go on to feature regularly in later debates about the limits of demonic power. Burchard included it because it cited Apostle Paul's injunction to shun heretics after their second admonition (*Titus* 3.10), which, he concluded, directed bishops to expel sortilegers (*sortilegi*) and workers of harmful magic (*malefici*) from their diocese. Other canons prescribed punishments for specific offenses. Canon 6, for example, specified that those who observe the pagan rites of divination or sortilege should be subject to five years of penance.[20]

Further canons in Burchard's work drew on the writings of the church fathers to distinguish between legitimate and illegitimate arts and offered a rationale for these decisions. Once again, several of Burchard's attributions were inaccurate. For example, he ascribed to Jerome the belief that it was acceptable to carry certain stones and herbs in order to ward off demons. "Jerome's" opinion was subsequently cited by later canonists, and it was later invoked by Christians to justify this practice.[21] More importantly, Burchard misattributed the materials in chapters 42–44, which contained some of the *Decretum*'s most sustained analysis of magic. He ascribed them to Augustine when they were in fact taken from book 8 of Isidore of Seville's *Etymologiae*. In these passages, Isidore condemned as superstitious a range of operative arts, including necromancy, hydromancy, pyromancy, geomancy, aeromancy, and all forms of astrological prediction. He also condemned, as had Augustine before him, "execrable remedies" proffered by physicians, clearly stating that ligatures and amulets produced effects only through the intervention of demons. Further individual canons condemned specific acts, such as the astrological practice of observing the Kalends of January.[22]

Later scholars such as Ivo of Chartres (1040–115) continued the work begun by Burchard. In his volume of canon law, Ivo reproduced many of the same materials as Burchard and, tellingly, repeated several of his misattributions. They included ascribing the canon *Episcopi* to the Council of Ancyra, the opinion that certain stones could ward off demons to Jerome, and sections of Isidore of Seville's *Etymologiae* to Augustine. Ivo also introduced a significant portion of extra material. Perhaps most notably, he directly cited passages from Augustine's *De doctrina christiana*. In these passages, Augustine had defined as superstitious the act of creating idols in order to worship them as God or for the purpose of making enquiries to or forming pacts with demons. Such pacts, he continued, were used in the magical arts. He specifically identified the arts of haruspicy and augury and the use of amulets and ligatures as examples of superstitious practices. Ivo then cited parts of Augustine's discussion of astrology, including passages in which he condemned as superstitious the activities of astrologers who attempted to use their knowledge of the stars to predict future human actions or the consequences of those actions.[23]

These collections of canon law were important for several reasons. First, they represented an attempt to collate and synthesize a disparate body of materials in order to produce a clear guide to ecclesiastical law. They also provided a set of guiding principles for considering the legitimacy of operative arts. While some of the attributions that they contained may have been misplaced, they put forward the consistent argument that magic and divination were superstitious in the technical sense that they required Christians to form pacts with demons to produce their effects. Furthermore, Augustine and Isidore of Seville were united in their condemnation of a wide range of specific operative arts. By giving such prominence to their opinions, these collections endorsed their conception of the acceptable limits for the discussion and practice of the operative arts. In effect, they prohibited virtually all of the magical arts, including numerous forms of divination—notably, predictive astrology—and various medical practices such as the fashioning of amulets or ligatures. If these canons were strictly applied, there seemed little scope for Christians to practice the operative arts. It is important not to overstate the influence of these early collections of canon law, however. Although they circulated relatively widely, they were not considered to be binding on the whole Church. Moreover, even if they had been accepted as definitive, the Church lacked the means to enforce them effectively. As we shall see, the Church's capacity to enforce canon law in a uniform manner only began to develop gradually from the mid-eleventh century when the papacy began to establish its supremacy.

The Papacy, Reform, and Orthodoxy

Despite, or perhaps because of, the political instability of the age, the tenth century witnessed the beginning of a series of local movements that sought in various ways to reform Christian society. To take one example, in order to protect clerics, ecclesiastical property, and members of the laity from conflict between members of the nobility, bishops in southern France cooperated with regional rulers to organize the Peace of God. While seeking to achieve specific practical ends, the movement also reflected an effort to reorganize society according to Christian principles, securing peace by distinguishing between just and unjust applications of violence.[24] The period also witnessed the beginning of a movement for religious reform. In 909, the duke of Aquitaine founded a new abbey at Cluny. Freed from secular intervention by papal protection, the community sought to establish a strict observance of the Benedictine Rule. The innovations at Cluny provided a model for monastic reform that spread rapidly; within a century, as many as a thousand reformed houses had been established. The ideals of Cluny inspired a series of similar reformist movements during the eleventh century. These included the establishment of further communities and orders that followed the Benedictine Rule, such as the Camaldolese hermits and cenobites and the Cistercian order. As I. S. Robinson has observed, however, this model of reform was conceived in a Pauline manner—that is, as a form of personal regeneration rather than an effort to reform the institutional church.[25]

By the beginning of the eleventh century, secular rulers such as the Capetians in France and the Salians and Saxons in the empire were beginning to restore their authority over and order within their realms. There was no immediate parallel in the Church. While the papacy retained a right established during the Carolingian period to invest metropolitans, who in turn appointed bishops, they had little influence over the affairs of local dioceses.[26] The situation began to change in the 1040s, when Emperor Henry III (1046–56) intervened in a dispute between factions of the Roman nobility vying for control over the papacy. Henry resolved the dispute by installing the German bishop, Suidger of Bamberg, as Pope Clement II (1046–47). This set a pattern for the next ten years, when each pope was a German appointed by the emperor. This series of popes, who combined the ambition to restore papal authority with a recognition of the need to respond to the growing clamor for renewal in the post-Carolingian Church, transformed the papacy. Their aspirations were informed by a radical new historical sensibility, which reconfigured the concept of reform: they believed that the Church had declined from an earlier state of perfection and that humans could restore it to its pristine condition. Henceforth, reform came to signify an attempt to regenerate the institutional Church and transform Christian society.[27]

The efforts to restore the institutional power of the papacy reached a new level during the pontificate of Gregory VII (1073–85). He declared that the papacy enjoyed legal supremacy over all Christians and that the clergy, hierarchically arranged under the pope, held power over all secular authorities. For Gregory, this meant that popes possessed the power to depose secular monarchs, including the emperor. He also insisted that all bishops were to be appointed by and owe direct obedience to the pope. By the end of the eleventh century, bishops were required to swear an oath of loyalty to the papacy when they were installed.[28] The papacy also began to assert a new legal role. Gregory VII declared that the papal court was now to act as the highest court in all Christendom. Furthermore, the papacy assumed the role of legislator and began to issue new laws, known as decretals. They were innovative because they did not simply amend or expand on existing laws but were in fact entirely new. Moreover, the papacy insisted that these laws had authority not only in the diocese of Rome but in Christendom as a whole. The papacy consolidated its legislative power by assuming the authority to convoke universal councils, a power previously reserved to the emperor.[29]

The advent of a new centralized legal structure necessitated a shared body of canon law. This was provided in the middle of the twelfth century when Gratian produced his *Concordantia discordantium canonum* (*Concordance of Discordant Canons*), more commonly known as the *Decretals*.[30] As Harold J. Berman has noted, this work "received almost immediate recognition as an authoritative statement of canon law." It became one of the core texts consulted and studied in the universities, and it was referred to by popes and councils. In his *Decretals*, Gratian made a further important contribution to the development of canon law by distinguishing between the body of laws that he was seeking to collate and systematize and the new papal decretals. The authority of the papal decretals was complemented by a further important innovation in legal theory developed by Gratian's pupil, Hugoccio, who maintained that papal decretals superseded any previously issued canons. Not only did this innovation further enhance papal authority, but it also contributed to the development of the idea of canon law as an evolving, amendable body of regulations rather than an established set of rules that required exposition. Canonists subsequently sought to systematize papal decretals, and compilations began to appear from the later twelfth century. This process was given new authority in 1234, when Gregory XI oversaw the creation of a definitive volume of these documents. This new volume and Gratian's compilation of existing canons formed the basis of a new system of canon law.[31]

Pastoral Reform and Social Discipline

Reforming popes also began to assert their duty to identify and eradicate error within Christian society and to oversee its replacement with correct doctrine and practice. In the age of Gregory VII, the rhetoric of reform was often invested with images of uprooting error and replanting with correct Christian doctrine and practice.[32] Since the papacy's power to directly implement reform was limited, it sought to create a structure of ecclesiastical governance that linked the papacy, via the episcopate, to local parishes to impose its authority. Bishops were at the center of this new system. Many of the decrees passed at the first two Lateran Councils, held in 1123 and 1139, defined bishops' responsibilities and their authority over pastoral care within their dioceses. Lateran I Canon 4, for example, stressed that a bishop must give his consent before anyone could have responsibility for the cure of souls. They were also encouraged to play a more active role in their dioceses, especially through preaching. Furthermore, bishops oversaw the creation of local churches with dedicated curates. The establishment of this parochial structure strengthened pastoral care, allowing communities to receive baptism, hear mass, and bury their dead. In principle, these changes also facilitated the reform of society by promoting clerical education and extending social and spiritual discipline at a local level.[33] Further Lateran decrees supported these measures—for example, by seeking to enforce celibacy among the priesthood and define acceptable degrees of consanguinity in marriage.[34]

The new parish-based system of social discipline created during the early reform period was subject to considerable local variation. The effectiveness of the reforms depended on the enthusiasm of the episcopate. In practice, episcopal oversight of priestly education was often limited, they trained few to preach effectively, and their efforts to encourage clerical celibacy frequently failed.[35] Social discipline was also applied unevenly. Since bishops retained the power to dispense justice, the pursuit of superstition and heresy was conducted at the diocesan level. The papacy had little control over these matters and during the first two Lateran councils did not treat them at any length.[36] Canon law was equally unhelpful. In his *Decretals*, Gratian had not included a specific discussion of magic. His only guidance on such matters was presented in a causa treating the excommunication of priests charged with magic. In this section, he reaffirmed Ivo and Burchard's conclusions that superstition could be punished by excommunication and cited familiar passages from Augustine and Isidore and the canon *Episcopi*.[37] In the absence of central direction, local synods defined heresy and superstition and prescribed punishments ranging from banishment and excommunication to corporeal punishment. In practice, the degree of toleration that the bishops were prepared to afford to heretics or those accused of superstition differed between dioceses. It is also important not to overstate the significance of these initiatives in the early period of reform. There does not appear to have been a concerted effort to eradicate heresy at a local level, and individual heretics often continued to live openly alongside their orthodox neighbors.[38]

By the later twelfth century, the papacy began to play a more active role in defining the parameters of orthodoxy and the administration of social discipline at the diocesan and parochial levels. These efforts began in earnest in 1179, when Alexander III convoked the Third Lateran Council.[39] Perhaps reflecting the failure of previous reform efforts, the fathers of Lateran III once more sought to address errors such as simony and enforce clerical celibacy. Further decrees indicate that the council fathers were increasingly concerned by the presence of non-Christians or deviant Christians living in their midst. Canon 26, for example, set out a series of principles to regulate the existence of Jews and Saracens living within Christian society. Although the decree underlined Jewish subservience to Christians, it did not remove the Augustinian principle of tolerance that had hitherto guided official attitudes.[40]

The Lateran decrees also demonstrated a new determination to tackle religious dissent. In some instances, the central issue was one of ecclesiastical authority. During the twelfth century, individuals and groups who sought to pursue new forms of religious life proliferated. The Waldensians were a notable example. They were followers of the perhaps mythical figure Peter Waldo (c. 1140–c. 1205), a cloth merchant from Lyon who, in around 1170, had renounced his wealth and begun a life of itinerant preaching. The council fathers examined the Waldensians' beliefs and judged them to be orthodox. In a development consistent with the reforms of the earlier councils that asserted the bishops' authority over the cure of souls, they nevertheless required the Waldensians to obtain episcopal licenses if they wished to continue preaching. When some of their number refused, they were condemned as heretics.[41]

The other significant group of heretics in this period were the Cathars. Catholics considered them to be adherents of an alternative dualist religion who sought to undermine the integrity of the orthodox faith.[42] Canon 27 demonstrated a new perception of the danger that the Cathars and other heretics posed. It acknowledged that although Catharism was not a new problem, its followers had previously kept their beliefs private. Now they preached them openly and by these means drew the "simple and weak" to their corrupted religion.[43] Their active proselytizing threatened not only their own souls but also those of orthodox Christians who lived near them. To separate the Cathars from the orthodox community, this canon declared that not only the heresy's adherents but also those who protected or traded with them would be placed under anathema. The council indicated that the laity too had their part to play in the struggle against heretics, by suggesting that they could "by arms protect the Christian people against them." The faithful were further incentivized to take direct action by the promise that those who fought the Cathars would be offered remission of sins. Although the Church never abandoned its efforts to use persuasion to convince dissenters to abandon their error, it was now prepared to sanction the application of violence to achieve these ends.[44]

The papacy evidently did not consider the measures enacted at Lateran III to be sufficient to meet the threat of heresy, for in 1184 Pope Lucius III promulgated the bull *Ab abolendam*. Inspired by developments in secular courts, the bull advocated the use of inquisitorial procedures for the investigation of heresy. This new procedure meant that instead of waiting for an accuser to bring forth charges, the investigator was required to examine rumors of criminal behavior. Mirroring this approach, *Ab abolendam* commanded that twice a year, bishops (or their appointed delegates) should inspect any parishes in their diocese rumored to harbor heresy. It further stipulated that clerics investigating heresy should work closely with secular authorities. Once they had identified heretics, they were to hand them over to the secular authorities for punishment. The bull nevertheless reflected the limits of papal power. Since the bull obliged local bishops to undertake inquisition, it was an assertion of papal power. But the provisions of the bull needed to be enacted at a local level, and its success depended on the compliance and enthusiasm of the episcopate.[45]

Papal efforts to create a new system of pastoral care within Christendom reached their climax during the pontificate of Innocent III (1198–1216). In 1215, he convoked the Fourth Lateran Council, which proposed a raft of measures designed to consolidate the papacy's control over the Church and establish clear rules to govern Christendom. Its opening canon provided a statement of orthodox Catholic belief, making it easier to demarcate the faith from heresy. Further canons introduced measures designed to assert the papacy's power to determine the orthodoxy of forms of religious life. Canon 13, for example, banned the institution of new religious orders and required anyone wishing to pursue a vocation in the regular clergy to join an existing order. It remained clear, however, that the papacy continued to envisage pastoral care operating through the existing diocesan and parochial structures. Canon 21, *Omnis utriusque sexus*, represented one of the council's most significant reforms. It required all Christians to confess their sins to their own parish priest at least once every year, enhancing the Church's surveillance of local communities.[46]

The council also set out further measures to control the relationship between Christians and non-Christians. Continuing the agenda laid out at Lateran III, Canons 67–70 sought to regulate the behavior of Jews resident within Christendom. Once again, this legislation did not introduce any major changes, but it did summarize and draw together preexisting measures.[47] The council's treatment of heresy was more innovative. Canon 3 reaffirmed the principle that those found guilty of heresy should be consigned to the secular arm for punishment. It continued that members of the clergy would first be "degraded from their orders" and their goods confiscated. Those suspected of heresy would be anathematized for a year, but if they had not made suitable satisfaction within that period, they too would be treated as heretics. The canon also proclaimed that any Catholic who took up the sword against heretics would be granted the same indulgences as those undertaking crusades against Muslims. This was not a theoretical power. In 1209, Innocent had launched a crusade against the Cathars in southern France, and military action continued in the region for a further twenty years.[48]

The Coming of the Friars

By the conclusion of Lateran IV, the Church had renewed its structures of ecclesiastical governance. Although the papacy had asserted its primacy over the episcopate, bishops enjoyed considerable autonomy to organize the secular clergy and supervise pastoral care within their dioceses. In the years that followed, this system was disrupted by the establishment of the orders of mendicant friars. They were founded as a result of the independent actions of two individuals. The first was Giovanni di Pietro di Bernardone (1181/82–1226), informally known as Francis, the son of an Umbrian silk merchant. Francis and a group of followers committed themselves to a life of apostolic poverty and began to preach in the towns of central Italy. The second was the Spanish nobleman and Augustinian canon Dominic de Guzmán (1170–1221). Having witnessed Catharism firsthand in southern France, he and a companion formed a group of mobile preachers to combat the heresy. Francis and Dominic each subsequently sought to transform their followers into new religious orders. Since Lateran IV had forbidden the foundation of new orders, Francis and Dominic were obliged to obtain papal permission for this course of action. It was granted in 1216 and 1223, respectively.[49]

Both orders of friars soon became closely associated with the papacy and its agenda. Francis and Dominic each envisaged their new orders playing an active part in the Christian community, tending especially to the populations of the rapidly expanding towns. This enabled them to extend pastoral care to areas where it might be insufficient and help implement the reforms of Lateran IV. The friars also drove campaigns against perceived enemies in Christian society—for example, the Jews. As Jeremy Cohen has shown, the thirteenth-century friars were instrumental in transforming the perception of rabbinic Judaism by arguing that the Jews had become, in effect, heretics to their own faith. The friars tended not to concern themselves with Jews directly (for example, by preaching in order to secure conversions) but instead sought to contain the perceived dangers posed by Judaizers or the circulation of the Talmud.[50] The friars also had a vital role to play in the fight against heresy. While Dominic had founded his order with the express intention of combating the Cathar heresy, the Franciscans too participated in this work. Individual friars such as Anthony of Padua (1195–1231), renowned for their skills as preachers, assisted in Innocent III's campaign against Catharism in southern France and northern Italy.[51]

Despite papal support, the early friars faced opposition from within the Church. Although he was personally well received by members of the ecclesiastical hierarchy, Francis's behavior and ministry recalled those of individuals such as Waldo of Lyon whose followers had been condemned as heretics. Moreover, the friars' activities created tensions within the Church. Their ministry disrupted the established order at the parochial level. Since they were directly accountable to the papacy, they were not subject to the oversight and scrutiny of local bishops. Members of the secular clergy also accused the friars of arrogating to themselves important parts of their pastoral role. The friars' preaching, they complained, was drawing the faithful away from parish churches; they were hearing confession, granting absolution, and offering burial. In short, the friars were emptying parish churches and depleting their finances.[52]

Questions about the friars' status and authority to carry out the cure of souls were brought to the fore during a further dispute that took place within the University of Paris. The friars entered the universities at an early stage of the development of their orders. Dominic intended his followers to combat the Cathar heresy primarily through preaching and disputation and recognized that to successfully perform this role it was essential that they possess an excellent education.[53] In 1217, he sent seven of his followers to Paris to enroll in the city's university, to seek vocations among their fellow students, and, more importantly, to partake of the education on offer. A mere two years later, they had established a priory with thirty friars. By 1224, the university was home to 120 Dominicans studying theology. By 1228, the Dominicans had created the order's first *studium generale*, which became known as St. Jacques. Members of the order subsequently founded priories in the university towns of Montpellier, Bologna, Oxford, and Cambridge.[54] Following the Dominicans' lead, the Franciscans also founded priories in Paris, Oxford, and Cambridge.[55]

The creation of mendicant *studia* demonstrated the new orders' ambition and intellectual capability, but it also rapidly created tensions within the university. These problems first surfaced during the Great Dispersion of 1229–31, when the university masters voted to suspend lectures. The friars chose to ignore the strike and continued to teach. The secular masters were further affronted when, during the strike, a Dominican friar, Roland of Cremona (1178–1259), took over one of the twelve teaching chairs in theology. The chair remained in Dominican hands even after Roland transferred to the newly established Dominican *studium* in Toulouse in 1230. The secular masters would later complain that the friars had acquired the chair by irregular means and that it should be returned. Shortly afterward, two secular theologians, John of St. Giles and Alexander of Hales, joined the Dominican and Franciscan orders, respectively. They brought with them their teaching chairs, which subsequently remained in the friars' possession. In 1231, Pope Gregory IX (1227–41) helped bring an end to the Great Dispersion by promulgating the bull *Parens scientiarum*. The bull also gave the university the authority to discipline any of its members—the mendicants included—who failed to respect its regulations.[56]

Tensions between the secular and mendicant members of the faculty of theology resurfaced in 1253, when the university again suspended teaching and the friars once more refused to participate. When this cessation ended, the university excommunicated the friars for refusing to obey academic statutes and expelled them from the consortium of masters. The friars were subsequently absolved by Pope Innocent IV (1243–54), but the university refused to allow them to resume their teaching duties unless they swore to abide by university statutes. The ensuing dispute between the mendicants and the seculars rapidly broadened in scope and escalated in acrimony. The seculars challenged not only the basis of the friars' ministry but also their right to exist. As these disputes developed, they exposed a series of ecclesiological questions that threatened to undermine the papacy's claims to supremacy. The secular theologian William of St. Amour (c. 1200–72) argued that Christ had invested each of the apostles with his authority, which they had in turn imparted to the episcopate. William therefore maintained that since episcopal and parochial jurisdiction derived directly from God, the bishop of Rome did not enjoy ex officio supremacy over other bishops and did not have the right to grant privileges—such as those enjoyed by the mendicants—that impinged on the existing rights of the secular clergy.[57]

The bitter disputes between the secular theologians of Paris and the mendicants continued throughout the third quarter of the thirteenth century, but the question of the mendicant order's legitimacy was concluded in 1275, at the Second Council of Lyon. In Decree 23, the council fathers reiterated Lateran IV's ban on the establishment of new religious orders. They also declared that existing unauthorized groups should cease to recruit new members and refrain from conducting the cure of souls by preaching, hearing confession, or offering burial. The canon nevertheless specifically made exceptions for the Dominican and Franciscan orders on account of the service they had rendered to the Church and for the Carmelite and Augustinian orders because their establishment antedated the Lateran decree. Although the theoretical discussions of the friars' right to exist had been settled and the debates on this matter had subsided, the question of the mendicants' authority at a local level remained unresolved. As we shall see in subsequent chapters, these disputes continued to reverberate in later centuries and affected the friars' activities, including the work of inquisition.[58]

Papal Inquisitors

By the early 1230s, the papacy had become increasingly dissatisfied with local efforts to prosecute heresy. It started to introduce significant changes to the practice of the inquisition of heresy. Although bishops would continue to be involved in this work, the papacy began to directly appoint investigators wielding the power of judge delegate to investigate allegations of deviance in particular regions. An individual exercising this power was referred to as *inquisitor hereticae pravitatis*—the inquisitor of heretical depravity.[59] At first, the effectiveness of this new initiative was limited by the papacy's inability to find sufficient candidates to discharge the office of inquisitor. On the one hand, the curia was relatively small, and the papacy could spare few of its officials for the task; on the other, many members of the local clergy lacked the education necessary to perform these duties. The papacy therefore began to draw its inquisitors from the Franciscan and, above all, the Dominican orders. Highly educated, owing direct obedience to the papacy, and possessed of an organizational structure dispersed throughout Europe, they proved ideal candidates.[60] There existed also a close connection between the work of inquisition and the Dominicans' founding ambition. Many members of the order saw no contradiction between their pastoral mission to preach against Catharism and the attempt to prosecute heretics. Indeed, as Christine Caldwell Ames has persuasively argued, many Dominicans considered the identification and suppression of heresy to be an integral part of their pastoral care.[61]

The papacy had assumed responsibility for appointing inquisitors to investigate and prosecute cases of heresy, but it did not create a formal, centrally directed organization. Individual inquisitors exercised the *officium inquisitionis hereticae pravitatis*—that is, the power of the office of the inquisition of heretical depravity—and possessed the authority to investigate offenses in their designated territory. The work of inquisition was therefore carried out by a series of independent officials who were answerable to the papacy. The creation of these new powers raised similar questions about the relative authority of the papacy and the episcopate to those created by papal dispensation to the mendicants to undertake the cure of souls within individual dioceses. As Richard Kieckhefer has stressed, the connections between individual tribunals were loose, and so it is no more appropriate to talk of an organized network of local inquisitions than of a single medieval institution. The structures supporting the work of inquisition, such as a system of archival records or a body of officials employed to undertake administrative tasks, also varied by location. Furthermore, most of the legislation defining the office of inquisitor was produced by local councils, which were, in turn, responding directly to the queries of local inquisitors. During the Council of Tarragona (1242), for example, the Dominican Friar and canon lawyer Raymond of Peñafort (c. 1175–1275) drafted a series of canons that set out the process of inquisition. Although latterly influential, they were not intended to provide a universal basis for inquisition. Consequently, the practice of inquisition remained highly localized, and its sophistication and effectiveness in a given area often depended on the enthusiasm of the incumbent inquisitor.[62]

Although the Inquisition did not exist as a single cohesive institution, inquisitorial practice was lent a degree of uniformity by the production of handbooks or manuals. First developed in legal, administrative, and financial contexts, the manual was a genre characteristic of the twelfth and thirteenth centuries. Clerics soon began to produce manuals to guide others in the work of preaching and confessing. From the 1240s, practicing inquisitors started to create manuals describing how to apply inquisitorial procedure to the prosecution of heresy. They continued to be produced in significant numbers until the early 1320s, when Bernard Gui and

Zanchino Ugolini produced two major works. After this date, the only significant medieval manual was produced by Nicholas Eymerich in 1376. As Lucy Sackville has argued, it is unhelpful to generalize about these manuals as they ranged in size, subject matter, and purpose. While some provided technical guidance on the process of running a trial, others offered a discussion of the nature of specific heresies; others still combined both kinds of information. It is also necessary to be wary of interpreting their development teleologically as a progressive, ever clearer elaboration of inquisitorial law and practice that culminated in the production of the "archetypal" works of Gui or Eymerich. Many manuals nevertheless contained consistent material, copied from one text to another. Several also cited substantial extracts from earlier writings such as the decrees of the Council of Tarragona.[63]

One of earliest examples of the new genre of inquisitorial manual was a relatively short work, *Ordo processus narbonensis,* which was produced in Carcassonne in the late 1240s. Prepared by Dominican friars in order to provide guidance to a confrère taking up the office of inquisitor, it provided clear legal formulas and outlined procedures to be followed during the prosecution of heresy. While the procedures described in *Ordo processus narbonensis* could have been directed against any type of heretic, much of its specific content reflected its compilers' preoccupations. Their primary concern was to provide information pertinent to the investigation of the activities of Cathars and Waldensians. Another inquisitorial handbook of this period, Rainerius Sacconi's *Summa de catharis et pauperibus de Lugduno* (1250), was a treatise offering relatively detailed descriptions of the titular heresies. Importantly for our purposes, neither text showed any concern for the investigation of superstitious magical belief or practice among otherwise orthodox Christians.[64]

The evidence from extant thirteenth-century handbooks therefore suggests that their authors chose not to write about magical belief and practices. The reasons for their silence are unclear. It is, however, unlikely that individual inquisitors had simply never encountered such beliefs or that they were entirely unconcerned by them. Early inquisitors' reticence about magic may have been informed by doubts about their authority to investigate such activities. It is evident that their powers in this area were unclear, for in 1258 several inquisitors wrote directly to Pope Alexander IV asking him to define the scope of their authority to investigate magic as inquisitors of heresy. The question raised theological as well as jurisdictional issues. Since the inquisitor was charged with the task of investigating acts of heresy, he did not necessarily enjoy the right to investigate cases of magic, which, as we have seen, Christian authorities had defined as superstitions. The pope was therefore required not only to define the scope of the Office of Inquisitor but also to pronounce on the theological status of magical practices.

Alexander responded in the bull *Quod super nonnullis* by declaring that "inquisitors of the pestilence of heresy, appointed by the apostolic see, should not intervene in cases of divination or sortilege, unless they manifestly savored of heresy; and nor should they punish such practices, but leave them to their judges for punishment." At face value, Alexander's judgment appeared to limit the inquisitor's powers. He expressly ruled that there was a swathe of activities that the inquisitor should not investigate but instead leave to the secular authorities. In this sense, the bull was upholding existing practice. Nevertheless, by stating that at least some activities could be considered heretical, Alexander had established an important legal precedent: that some forms of magic and divination were heretical. This judgment therefore appeared to contradict the principles established in canon law that magic and divination were illicit because they were superstitious.[65]

Alexander's judgment was of fundamental importance because it laid the legal basis for the future inquisition of magic. As we shall see, however, it was also notable because it left these powers undefined. Alexander did not offer a clear definition of what constituted "manifest heresy." He may, for example, have been referring to forms of ritual magic that required making explicit contact with demons. Even if Alexander had sought to determine that operative arts that involved making compacts with demons were heresies, his letter did not specify which particular examples he had in mind. This was particularly significant because, as we have seen, many operative arts—for instance, astrology and geomancy—functioned by contested means. Alexander's letter did not identify any specific criteria for determining whether any of these arts required demonic involvement. The precise forms of sortilege or divination that fell under this category were therefore open to interpretation. This meant that although inquisitors did now enjoy the right to prosecute some forms of heretical magic, the precise limits of their authority remained undefined.

Conclusion

By the middle of the thirteenth century, the papacy had reinvented itself as an institution. Having asserted its supremacy over the entire western Church, the papacy assumed the right to act as the arbiter of the orthodox faith, which it defined by using institutions such as universal councils. It had also begun to create hierarchically ordered structures to administer pastoral care, which continued to be under the direction of the episcopate. The other significant innovation of this period was the creation of new institutions that owed obedience to the papacy, which were designed to encourage orthodoxy and where necessary eliminate heresy, whether by means of persuasion or coercion. These twin developments formed the context for a transformation in thinking about the legitimacy of magical or operative arts. For most of the period considered in this chapter, these arts remained a marginal concern for the papacy, yet by 1258 inquisitors of heretical depravity had been given the power to investigate a limited range of magical offenses. It remained unclear, however, which forms of magic were to be considered heretical or even which criteria should be used to make such decisions. As we shall see in the next chapter, two members of the Dominican order, Albertus Magnus and Thomas Aquinas, wrote extensively on these matters, providing new sets of censorial criteria that partly expanded and partly contradicted those established in canon law.

✷ 2 ✷

The Dominican Order and the Construction of Orthodox Magic

Introduction

By 1258 the papacy had invested inquisitors with the authority to investigate forms of magic that manifestly savored of heresy. It was, however, unclear which of the operative arts fell into this category. The task of making these decisions was made harder rather than easier by the intellectual developments that occurred between 1000 and 1300. During the eleventh and twelfth centuries, canonists such as Burchard of Worms, Ivo of Chartres, and Gratian had rediscovered a series of judgments that helped determine the legitimacy of the operative or magical arts. Endorsing the opinions of church fathers such as Augustine and Isidore of Seville, they had argued that the overwhelming majority of these arts were superstitious and should be avoided by Christians. Canon law, in effect, imposed an outright ban on their practice. This situation started to change from the mid-thirteenth century when several Scholastic thinkers began to reconsider the boundaries between legitimate and illegitimate ideas and practices. They included two illustrious members of the Dominican order, Albertus Magnus and Thomas Aquinas. In a marked change from the arguments produced by the earlier canonists, both friars were prepared to accept that it was permissible for Christians to practice certain operative arts. More importantly, they reconceptualized the criteria that should be used to determine the legitimacy of a given art, by arguing that arts with demonstrably natural causes were legitimate. As we shall see, to determine which of these arts were naturally possible and which could produce their effects only with the assistance of demons, they drew on not only theology but also pagan philosophical knowledge of the natural order.

The Friars and Aristotle's Libri naturales

The orders of mendicant friars had been founded to pursue a pastoral ministry, primarily within the urban centers of the high medieval period. The need to provide highly educated preachers who could educate the masses and argue against those who held erroneous views ensured that education played a central role in these orders. As Roger French and Andrew Cunningham have argued, for differing reasons the Franciscan and Dominican orders became especially interested in pagan knowledge of the natural order and especially Aristotle's *Libri naturales*. While early Franciscan intellectuals such as Bonaventure (1221–74) used the study of nature as a means to contemplate God and his creation, Dominicans such as Albertus Magnus and Thomas Aquinas used pagan wisdom to buttress the faith and enter polemical battles with opponents such as the Cathars and, as we shall see, philosophers and practitioners of the operative arts. Using these resources, the orders of friars gradually created distinctive forms of natural philosophy, which were explicitly designed to create demonstrated knowledge about God's creation. Controversially, they used Aristotle's works for these purposes.[1]

In the first half of the thirteenth century, Aristotle's thought had an ambiguous status in the Latin West. While his *Libri naturales* and *Metaphysics* had been freely accepted at the University of Oxford, the teaching of these works at the University of Paris had been effectively banned by Gregory IX (1227–41) in the bull *Parens scientiarum* (1231). There were good reasons for Christians to be wary of Aristotle's conception of the natural world. At its heart lay an attempt to comprehend the regular, predictable laws that governed the everyday operation of the cosmos. Although this powerful interpretative framework offered a means to understand and predict effects in the natural world that was useful to practical human activities such as learned physick, Christians also acknowledged that, in several key respects, Aristotle's cosmology was antagonistic to their faith. First, Aristotle's cosmos was eternal; it had neither beginning nor end in time. Such an idea flatly contradicted the Christian belief that God had created and would one day destroy the world. Second, Aristotle maintained that the whole cosmos was given motion by the prime mover, a deity sublimely unaware of the world that depended on him. All effects in the terrestrial realm could be explained by tracing a chain of causation that ultimately led back to the prime mover. Many medieval interpreters of Aristotle's work believed that he had intended these effects to include not only natural phenomena but also human actions. Understood in this way, Aristotle's cosmology was inherently deterministic and therefore in conflict with Christian belief in free will.[2]

Aristotle's naturalist cosmology also had indirect implications for Christian orthodoxy. His conception of the deity as prime mover was far removed from the Judeo-Christian idea of a providential God who not only was deeply concerned with his creation but also directly intervened in it by performing miracles. Since the prime mover's influence on the cosmos was indirect, he did not—indeed, could not—act directly to alter the regular laws of natural causation. In other words, he did not perform miracles. Furthermore, in a strict Aristotelian account of the cosmos, it was not possible to identify demons or—in a Judeo-Christian context—angels as a potential cause for any observable phenomenon. This aspect of Aristotle's cosmology again conflicted with Christian belief. As Augustine had observed in *De civitate Dei,* it was impossible for a Christian to deny either the existence of demons or their ability to work effects in the world, as each was attested in scripture.[3] Consequently, the possibility of demonic magic, no less than miracles, was an essential feature of orthodox belief. A thoroughgoing Aristotelian naturalism was therefore corrosive to Christianity, precisely because it threatened to remove not only divine (that is, supernatural) but also angelic and demonic (that is, preternatural) causation from cosmological theory.

Although Aristotle's *Libri naturales* remained the subject of controversy in Paris, they came to play an increasingly significant role in the education of Dominican friars. At a relatively early stage of their engagement with the Cathars, some Dominicans had realized that the heretics were using Aristotelian arguments to defend their belief that the world was an evil creation. To meet these arguments directly, they too began to study Aristotle's works. It is highly likely that, despite the ban on teaching Aristotle's *Libri naturales* in the University of Paris's faculty of arts, the Dominicans continued to read these works in their own houses.[4] After the death of Pope Gregory IX in 1241, however, Aristotle's *Libri naturales* began to be taught openly. These efforts were supported by William of Auvergne (1180/90–1249), the bishop of Paris and chancellor of the university. William was an accomplished theologian who also possessed a sophisticated understanding of Aristotle. He also shared the Dominicans' intense concern about the Cathar heresy and had developed similar ideas about how it could be challenged through the use of Aristotle's philosophy. With his support, Aristotle's ideas became gradually incorporated into the teaching at Paris, and by 1255 lectures on his works had become a mandatory part of the arts curriculum.[5]

Within the Dominican order, Aristotle's *Libri naturales* were taught primarily in the order's *studia* in Paris, Bologna, and Oxford. The teaching of philosophy was not confined to the universities, however. Since 1220, each priory had been required to establish a school offering theological instruction. Throughout the 1240s and 1250s, two of the Dominican provinces had allowed their schools to teach Aristotelian philosophy. Although the

Dominicans were early proponents of the study and use of Aristotelian philosophy, the training in this subject that they offered in their schools during the first thirty years of the order's existence was far from comprehensive. Often working collaboratively, early Dominicans collated selections from pagan philosophers including Aristotle that they considered useful for teaching within the order or preaching to the laity. These compendia of pagan wisdom, often presented as paraphrases, offered some insight into these complex texts and their commentaries but not a systematic understanding of them.[6]

Philosophy and Magic in the Thought of Albertus Magnus

The Dominican order's approach to the interpretation, exposition, and teaching of Aristotle's texts was transformed by the work of Albertus of Lauingen (1200–1280). He was recruited into the Dominican order in 1223, while studying the liberal arts at the University of Padua. He was a gifted scholar, and his contemporaries considered his intellectual feats so impressive that within his own lifetime they referred to him as "Magnus" or "the Great." In the 1240s, he was sent to the Dominican *studium* in Paris to take up a position as a bachelor of the sentences. On completing his degree in 1242, he assumed one of the two chairs in theology held by the Dominican order in the University of Paris. While performing these duties, he met a young Dominican, Thomas Aquinas (1225–74), who had arrived at St. Jacques in 1245. In 1248, Albertus, accompanied by Aquinas, left Paris to establish a new *studium* in Cologne. While there, he began to produce paraphrases of and commentaries on the Aristotelian corpus; these included studies of works that would now be considered "pseudo-Aristotelian." His labors represented a milestone not only in the development of the intellectual history of the Dominican order but also in that of the Latin West. As David Lindberg has observed, his work represented the "first attempt to provide a comprehensive interpretation of Aristotle's philosophy in Western Christendom."[7]

Albertus also produced several original philosophical works. Some of these books—for instance, *De mineralibus,* which dealt with the classification of types of stones and their properties—considered topics that Aristotle appeared not to have addressed. By writing this and other texts, Albertus may have been attempting to fill perceived gaps in the Aristotelian corpus. It is certainly clear that he used Aristotelian works as models for arranging and analyzing his source material.[8] Historians also continue to debate whether Albertus was the author of works such as *Speculum astronomiae* and *De secretibus mulierum,* which treated various forms of magical activity. Irrespective of modern debates about their authorship, many contemporaries did believe that Albertus wrote these texts, and many Christians continued to attribute them to him until at least the sixteenth century. It is also likely that Albertus wrote at least some, if not all, of the material that they contained. There are certainly strong thematic similarities between these texts and those that Albertus indisputably wrote. In texts such as *De physica* and *De mineralibus*, for example, Albertus considered questions concerning the nature of astral influence and whether humans could harness this power in a manner similar to the author of the *Speculum*.[9]

Albertus did not comment directly on his reasons for producing this vast series of commentaries. He did, however, provide some indication of his motives in the introduction to his commentary on *De physica* (1249). In the first instance, he stated that he wanted to glorify God by achieving a better understanding of his creation. Studying Aristotle's natural works could, he maintained, help him achieve this aim. He also indicated that his commentary on *De physica* served a specific purpose within his order. He had, he recalled, been moved to write this particular commentary in order to satisfy his confrères' desire to better understand the content of Aristotle's ideas. Pursuing this ambition, Albertus sought to replace the earlier system of teaching selected extracts of Aristotle's thought with a comprehensive yet accessible introduction to his work. He therefore presented his commentary on *De physica* in a form that would make it suitable for teaching in the Dominican *studia*. It provided students with an introduction to philosophy useful as a preparation not only for the further study of Scholastic theology but also for polemical battle with heretics such as the Cathars.[10]

Like many of his fellow Scholastics, Albertus was conscious of the dangers posed by philosophy in general and Aristotle in particular. Albertus gave one of the clearest indications of his ideas about the correct relationship between philosophy and theology in an early work, his commentary on Peter Lombard's *Libri quattuor sententiarum*. In a section treating the Hexameron, he discussed the nature of light and posed the question of whether it was a body or a form of body. To settle the matter, he consulted both philosophical and theological authorities. He could marshal an array of opinions garnered from the works of philosophers such as Aristotle and Avicenna to defend the position that light was not a body, but he was confronted by a significant problem: Augustine had seemingly defended the opposite opinion. Albertus was therefore faced with the problem of resolving a conflict between two different kinds of authority: one pagan and philosophical, the other Christian and theological. In this context, one might have expected Albertus to defer to the views of Augustine, who had written extensively on the use of pagan philosophy in the service of the faith. Albertus concluded, however, that "in those matters which concern the faith and morals, Augustine is more to be believed than the philosophers, if they disagree. But if medicine is spoken of, I will believe Galen or Hippocrates more: and if natural things are discussed, I believe Aristotle more than any other expert in natural things."[11]

Albertus's remarks provided an important insight into the criteria that he developed to assess the legitimacy of pagan knowledge. He was prepared to grant philosophy a large degree of autonomy. In those instances when a Christian was seeking to understand God's creation, he suggested that the opinion of a pagan philosopher could be given precedence over that of a Christian authority. This remained true even when that authority was a church father of the stature of Augustine and even when a Christian was undertaking a specifically theological task such as exegesis. Albertus was not, however, arguing that pagan philosophy should be considered more authoritative than scripture on natural matters. Instead, he was suggesting that if there were a contested reading of scripture, one where the point at issue concerned a Christian's understanding of the natural order, then the opinion of a pagan philosophical authority should be preferred over that of a Christian theological one. He did, however, reject philosophical accounts of the natural world when they unambiguously conflicted with Christian belief. Consequently, he was not prepared to accept some specific Aristotelian ideas—for instance, the eternity of the cosmos or the immortality of the soul. To give a philosophical account of these subjects that was in accordance with the faith, he used the writings of other philosophers.[12]

Albertus's willingness to challenge acknowledged Christian authorities and to give a more receptive audience to pagan claims about the natural world also informed his approach toward the operative arts. As we saw in the last chapter, in their compendia of canon law Burchard of Worms, Ivo of Chartres, and Gratian had endorsed Augustine and Isidore of Seville's rejection of these arts. Like a number of his contemporaries—for instance, the eminent Franciscan Roger Bacon (1214–92)—Albertus believed that it was acceptable for Christians to practice at least some of them. While he rejected arts that overtly appealed to angels or demons, he maintained that arts such as the construction of talismans might be acceptable. Unlike Bacon, who developed the concept of *Scientia experimentalis*, Albertus did not offer a systematic rationale for his engagement with the operative arts.[13] It is, however, possible to reconstruct aspects of his thought on these matters. Albertus was conscious of the fact that several of these arts were potentially superstitious and that it was therefore necessary to distinguish those that were legitimate from those that were not. His approach to this task was underpinned by his belief that any set of practices that possessed demonstrably natural causes were, by definition, not demonic and so should be considered legitimate. If this principle appeared simple, it was only superficially so. To be useful, it required Albertus to establish how nature operated and the extent to which those operations could be manipulated by humans. Moreover, he needed to show how these decisions should be made.

As we have seen, Albertus believed that pagan philosophers (rather than Christian authorities) provided the most plausible accounts of nature and, by extension, of the effects that humans could produce by manipulating natural qualities. Determining the legitimacy of a particular art therefore required him to decide which of the available philosophies offered a true account of the manner in which the cosmos functioned. Following the principles outlined in his commentary on the *Sentences*, Albertus invoked philosophical rather than theological authorities to arbitrate between disputed accounts of the natural world. In practice, this meant that he used rational discussion—sometimes bolstered by empirical observation—to choose between them. To take one example in *De mineralibus*, a text written between 1250 and 1254, Albertus used philosophical accounts of nature to justify several claims about the properties of stones. Many, he noted, possessed wondrous properties, the causes of which were occult or hidden, albeit naturally occurring. From the standpoint of Christian tradition, this clear assertion of the reality of occult qualities was relatively unobjectionable. Augustine, for example, had argued that it was possible for some natural substances to possess wondrous properties and even specifically remarked on the hidden properties of stones. More recently, eleventh-century canonists, such as Gratian, had ascribed to Jerome the opinion that certain stones could ward off demons. Albertus was prepared to give far more weight than either Augustine or the canonists to the accounts of nature provided by philosophers, however. Drawing on a range of pagan sources, he articulated a general theory to explain how stones could naturally acquire the various properties that he ascribed to them.[14]

Albertus's deference to philosophical authority also meant that he was prepared to accept the natural possibility of a far wider range of effects and, as a result, to accept the legitimacy of a number of magical arts that Augustine and the canonists had condemned. For example, in specific circumstances he was prepared to accept the orthodoxy of the practice of constructing talismans or making inscriptions.[15] Albertus certainly acknowledged that certain images or sigils were explicitly necromantic, in the sense that they were specifically designed to solicit the assistance of demons, and that Christians were therefore forbidden to make or use them. He nevertheless maintained that other images were in fact acceptable. As far as Albertus was concerned, if Christians could successfully distinguish licit from illicit images, then there was no reason why they could not use them. There remained, however, the problem of determining which were acceptable and which were not. Albertus observed that this delicate task had been rendered almost impossible by the comparative ignorance of contemporary Christians. "Few indeed," he observed, "understand the writing of the sages of antiquity regarding sigils of stones, and it cannot be known unless the sciences of astronomy, magic and necromancy are all simultaneously understood."[16] Albertus implied that with his newly acquired understanding of these sciences, he could help his fellow Christians distinguish acceptable from unacceptable sigils.

Albertus proceeded by describing the various types of image that a Christian might encounter, noting that they could be produced both by nature and by human artifice. For Albertus, both kinds of image could possess occult qualities.[17] After first describing the various kinds of naturally occurring images, Albertus proceeded to discuss images made by human artifice, specifically those produced by the practice of carving gems. He explained that this art had been developed and perfected by ancient magicians such as Magor of Greece, German of Babylon, and Hermes of Egypt and later illuminated by the likes of Ptolemy and Geber of Seville. Thebit ben Corat, he added, had subsequently provided a thorough account of these ideas. Albertus stressed that this art worked by natural means, noting that "it is, however, the principle of this science that all things whatsoever that are made by either nature or art, are first moved by the celestial virtues."[18] Some naturally occurring images produced by the stars were, he continued, imbued with these powers. He then explained how a human might artificially produce similar images. According to Albertus, the production of talismans was itself guided by the stars. Celestial virtues had the power to incite a person to make something—for example, a carved gem—and to direct how he or she fashioned it. Recognizing that this account of the stars' power to influence human actions potentially contradicted the principle of free will, Albertus noted that "there is in man a twofold principle of action."[19] The first principle, nature, was controlled by the stars, while the second, the will, was free. Although the will could resist nature's influence, if it failed to do so "the will then [began] to be inclined to the motions and figures of the stars."[20]

Having established that the stars could indeed influence the formation of both natural and artificial figures in gems, Albertus turned to consider how they might also impart to the figures their causal influence. He noted that the philosophers he had mentioned at the beginning of his chapter had already proved that the figures of the stars were principal figures, which had precedence over the figures of all things made by nature and by art. The configuration of the stars therefore exerted a causal influence on every figure produced on Earth, whether by natural or artificial means. Some of the power of that configuration was poured into the work of nature or art. For this reason, he noted, Ptolemy recommended that even simple actions should be performed only when the positions of the stars had been considered. He further maintained that geomancy, a divinatory art, could also be effective, but only if the geomancer restricted himself to considering those figures produced by the stars. These principles also informed the art of carving images on gems or metals. The masters of this art recommended that the carving should be carried out at precisely specified times, when the heavenly forces could exert the strongest possible influence. Although Albertus did not make any specific claims, he added that such masters "worked wonders by means of such images."[21]

In *De mineralibus*, Albertus also argued for the possibility of several alchemical practices, most notably chrysopoeia, the art of transmuting metals.[22] Defending alchemy was a less controversial undertaking than advocating the practice of carving gems. As William Newman has observed, at this time alchemy was rarely, if ever, associated with necromancy. On the contrary, he noted, contemporaries regarded it as "a perfectly reasonable and sober offshoot of Aristotle's matter theory."[23] Discussion of alchemy therefore often concerned its physical possibility rather than its orthodoxy, and it tended to be conducted by means of philosophical examination. During the thirteenth century, debates about the possibility of chrysopoeia centered on arguments originally formulated by the Persian philosopher Avicenna in *The Book of the Remedies*, which medieval Christians frequently misattributed to Aristotle. Critics of alchemy seized especially on one of his arguments, which was rendered in Latin as "*Sciant artifices, species transformari non posse*" ("Artificers should know that species cannot be transformed"). Using this argument, they maintained that it was impossible for humans to know the underlying properties of a specific metal. Such ignorance of a metal's properties meant they could not manipulate them, with the corollary that they could not successfully transform one metal into another. Albertus dismissed or, perhaps more accurately, argued around *Sciant artifices*. By interpreting the Latin *species* to mean "specific forms," he was able to draw on existing Scholastic theory to argue for the physical possibility of transmutation.[24]

As we have seen, Albertus was also believed to be the author of further texts treating various forms of magic, such as the *Speculum astronomiae*.

Even if Albertus did not write the *Speculum astronomiae,* its author defended ideas and practices similar to those defended in *De mineralibus*—for example, the production of astrological talismans and images to create effects. The author also considered whether it was possible to use the stars to predict the future. Church fathers such as Augustine and Isidore of Seville had declared that this art was superstitious, a view that was once again shared by the canonists. The author of the *Speculum* nevertheless argued that at least some forms of astrological prediction were in fact possible. One of the text's most important and influential sections contained a defense of the principle that the stars could exert a natural influence over the human body and, in turn, affect behavior. The text suggested that by understanding the correlation between astral influence and human behaviors, it would be possible to predict future human actions. The author of the *Speculum* took steps to ensure that his account of astral influence was not deterministic by introducing an idea similar to the "twofold principle of action" presented in Albertus's *De mineralibus*. He argued that although the stars could act on the body, influencing character and perhaps inclining an individual toward predictable actions, those inclinations could ultimately be overruled by the human will. The stars' influence was therefore not deterministic.[25]

Philosophy and Magic in the Thomist Synthesis

In 1252, Aquinas returned to Paris from Cologne. There, he continued his studies as a bachelor of the sentences and completed *Scriptum super sententiis,* his commentary on Peter Lombard's *Libri quattuor sententiarum.* In this early work, Aquinas began to develop a radical new means to manage the relationship between philosophy and the faith. It was informed by Augustine's concept of a unity of knowledge, which maintained that the practitioners of theology and philosophy should be able to arrive at complementary conclusions. Aquinas bolstered this idea by using arguments derived from the Aristotelian classification of knowledge. He maintained that both philosophy and Scholastic theology were sciences in the Aristotelian sense—that is, that they employed human reason to create demonstrated knowledge or scientia. There was, he continued, a hierarchy among these sciences. The starting point for a natural philosophical demonstration was information about God's creation that humans had gathered for themselves via their senses. Aquinas argued that since theological demonstrations took for their starting principles knowledge derived from scripture, which was revealed directly by God, theology possessed a higher degree of certainty than philosophy. Consequently, the latter discipline was subalternate to the former.[26]

Aquinas developed this argument by suggesting that while philosophers could use their methods to draw conclusions about God's creation, they could not contradict the conclusions established in the higher discipline of theology. It would, for example, be impossible to conclude on the basis of philosophical arguments that the world was eternal. Theologians therefore possessed the right to determine whether any philosophical argument was true, but philosophers could not challenge the conclusions drawn in theology. This meant that although Aquinas's synthesis gave philosophy a large degree of autonomy, it also provided a clear rationale for theologians to determine which philosophical ideas were false. In formulating these principles, Aquinas was not seeking to impose alien theological concepts on nascent forms of science, nor was he advocating the practice of a "bad" or even an "immature" science; he was seeking to articulate and regulate the boundaries of a highly sophisticated natural philosophy.[27]

Aquinas's synthesis, which placed clear restrictions on the practice of philosophy, also had significant implications for contemporaries' understanding and discussion of magic. In book 4 of *Scriptum super sententiis*, Aquinas noted that some individuals had denied the reality of *maleficium* on the basis that all wondrous effects had natural, albeit occult, causes. Although Aquinas did not state who had offered this opinion, this was a conclusion that could be drawn from a strict interpretation of Aristotle's philosophy, as we have seen. Aquinas dismissed all such arguments by stating that disputing the possibility of *maleficium* on these grounds would be tantamount to denying the existence of demons. Furthermore, he continued, by holding such an opinion, "these people repudiate the true faith, according to which Angels fell from heaven, and we believe them to be demons, and that from the subtlety of their nature they can do many things that we cannot; and therefore those who induce demons for the purpose of doing such things, are called *malefici*."[28] With these words, Aquinas denied the truth of strictly naturalist philosophical accounts of the cosmos on the ground that they conflicted with theological precepts. Within his synthesis of knowledge, defending a cosmology that denied demonic agency would be no more acceptable than advocating one that implied the world's eternity.

Aquinas's system of subalternation therefore imposed a series of requirements on philosophers. It was restrictive, in the sense that it was intended to allow theologians to delimit the parameters and review the content of the knowledge that philosophers made about the natural world.

In other respects, Aquinas's synthesis was productive. It helped philosophers generate knowledge of God's creation and all that it contained. When offering a true account of the cosmos, the philosopher was obliged to accept—and he certainly could not deny outright—the possibility of preternatural or supernatural causation. Importantly, however, Aquinas's synthesis obliged a philosopher not to always ascribe an event to these orders of causation but merely to consider them as possibilities. The challenge facing natural philosophers working within this system was to understand the boundaries of natural causation in a manner that accorded with the truth of theology. Operating within these parameters, they could offer a true account of physical effects in the cosmos and establish a means to determine the legitimacy of a particular operative art. On the one hand, Aquinas's synthesis required philosophers to establish which effects were genuinely natural and whether humans could manipulate them. On the other, it also obliged philosophers and theologians to articulate their understanding of what lay beyond the ordinary working of nature.

To determine nature's limits in the manner envisaged by Aquinas required theologians and philosophers to make detailed reflections on the extent of the power of demons and angels and their ability to work effects within the created order. Aquinas considered these matters in book 2 of the *Scriptum super sententiis*. In this context, reflection on the natural order offered a valuable heuristic. One of the questions he posed was whether it was possible for demons to induce true effects in bodily materials. He concluded that, like humans, demons could induce some changes in material but that they could do so only by means of art—that is, by skillful manipulation of the created order. To define the limits of what could be achieved by art, he discussed Avicenna's *Sciant artifices*, which considered whether the transmutation of metals was possible. Unlike Albertus, however, Aquinas sided with Avicenna, concluding that neither humans nor demons could genuinely effect such a transmutation. In this instance, Aquinas used the argument that chrysopoeia was naturally impossible as a means to demonstrate the limits of demonic power.[29]

In 1256, Aquinas was appointed to the regent chair of theology in Paris. He remained there until he returned to Italy in 1259. This was to be a particularly productive period of his life. In addition to completing several smaller works, Aquinas produced the *Summa contra gentiles* and began work on his *Summa theologica*.[30] He wrote each work to fulfill a specific purpose. While Aquinas conceived the *Summa theologica* as an introduction to all of theology, the *Summa contra gentiles* was, in French and Cunningham's words, "intended as a guide to Dominican missionaries spreading the word of true Catholic orthodoxy" among the infidels, apostates, and heretics.[31] In other words, it was to be used by friars in their pastoral ministry. Divided into four books that respectively treated God, his creation, the working of providence, and salvation, it offered preachers a guide to heterodox ideas that they might encounter during that ministry and presented arguments derived from natural reason that they could use to refute them.

Since the deployment of philosophy was central to the project described in the *Summa contra gentiles*, Aquinas set out not only its uses but also its limitations. Aquinas explained that the truth relating to God was twofold. The first class of truths about God—for instance, the fact that he is triune—humans simply could not learn through reason. Such truths could only be revealed by God, and humans should hold them as a matter of faith. Although full understanding of these truths would always exceed humans' natural capabilities, they could explore them to a limited extent. Such activity was properly the domain of theology. Philosophy could provide some probable arguments in support of these truths, but demonstrations of this kind could only be used to buttress the faith of those who already believed. The manifest inadequacy of these mere human arguments meant that they should not be used for the purpose of proselytizing. Aquinas nevertheless maintained that a second class of truths about God—the fact of his existence, for example—could be demonstrated by natural reason alone and that humans should use philosophy to both demonstrate these truths and repudiate contradictory arguments.[32] Pursuing this ambition, Aquinas developed arguments designed to rebut specific heretical propositions, such as the Cathars' contention that the world was the creation of an evil principle.[33] Several of Aquinas's arguments were addressed to opponents within the Catholic Church. For instance, he took the opportunity to counter ideas put forward by Aristotelian philosophers. Using philosophical arguments, he demonstrated the truth of the immortality of the individual human soul and the creation of the earth and disposed of contradictory arguments.[34]

Some of the arguments that Aquinas developed were directly relevant to the friars' efforts to combat superstition. In part 3 of the *Summa contra gentiles*, he considered the nature of divine providence. Several of the questions treated ideas that were directly relevant to the theoretical discussion of magic. Aquinas set out arguments defending various propositions about God's creation and the manner in which it functioned, his powers to work effects within that creation, and those that he gave to his creatures. During this discussion, Aquinas defined miracles as effects that either contradicted nature or worked through nature in an unusual manner. They were, in short, feats that God alone could accomplish. Turning to consider the powers of angels and demons, he reiterated his belief that they produced wonders only through the manipulation of natural causes. Defining the boundaries of the natural order, he argued that the stars could make no direct impression on human intellects, meaning that they could not determine human actions.[35]

In question 105, Aquinas addressed the more specific question of how magicians worked their effects, which, he observed, "can indeed be easily considered if we pay close attention to the means of working."[36] Aquinas continued by discussing whether humans could work effects through signs, such as spoken or written words and other symbols. The first category referred to forms of uttered spells or incantations, while the second designated the production of material objects such as talismans. Both forms of signification, he concluded, ultimately had no inherent efficacy but instead appealed to an addressee—that is, a demon who produced the desired effect. He also noted that some might object that it was possible to produce astrological talismans. Discussing this point, Aquinas concluded that although bodies might be affected by celestial bodies, the act of inscribing them made no difference to their capacity to work effects. While Aquinas appeared to reject the act of making any talismans, including astrological ones, he nevertheless introduced one final counterargument, which appeared to suggest that products of human artifice could receive celestial influences. Although it seemed to contradict his earlier views, curiously, Aquinas did not rebut this argument. As Darrel Rutkin has observed, Aquinas "seems to leave open a gaping loophole here to justify the production of natural talismans of the type that Albertus Magnus and the *Speculum astronomiae* defended as legitimate."[37]

Aquinas and the Teaching of Philosophy

By the mid-thirteenth century, Aristotle's philosophy enjoyed an established position in western European intellectual life. In 1256, a year after Aristotle's philosophy became an official part of the arts curriculum in Paris, Albertus and Aquinas participated in a general chapter that established a *ratio studiorum*, which enshrined the teaching of Aristotle's philosophy within the Dominican pedagogical system. These were not inevitable developments. As we have seen, during the 1240s and early 1250s Dominican friars such as Albertus Magnus and Thomas Aquinas had played a significant role in promoting the use of Aristotle's thought both within their order and in the universities, and in developing strategies to ameliorate the risks that his ideas posed to the faith. Although the teaching of Aristotle's thought was now officially sanctioned in the universities and the Dominican order, some theologians—including some drawn from the mendicant orders—remained wary of the implications of a number of his ideas.[38]

Debates over the legitimacy of Aristotle's *Libri naturales* were reignited during the 1260s, when some arts masters at the University of Paris began to use a new means to teach and discuss philosophy. Their approach is often somewhat misleadingly referred to as "Averroism." Following Fernand Van Steenberghen, I shall refer to it here as Radical Aristotelianism. Its proponents included such philosophers as Siger of Brabant (1240–84) and Boethius of Dacia. They argued that the philosopher's role was to teach and interpret Aristotle's thought in a manner that they considered accurate, irrespective of its implications for the Christian faith. Seeking to insulate themselves from the charge of promoting false teachings, they maintained that it was possible to distinguish "necessary" conclusions—those constructed by arguing purely as philosophers—from the "true" conclusions that they held as Christians. When speaking as philosophers, they could, for example, hold that rational argument demonstrates the necessity of the world's eternity. When speaking as Christians, however, they would declare the absolute falsity of this position and instead assert the truth that the world was created and would eventually be destroyed by God. By expressing their ideas in this manner, the Radical Aristotelians were not seeking to undermine the faith. On the contrary, they were acknowledging the limitations of mere human knowledge in comparison to divinely revealed truth. They were, in this sense, radical fideists.[39]

The Radical Aristotelians' pious intentions notwithstanding, their ideas posed several challenges in the late medieval universities. First, they made it possible to publicly teach ideas that appeared to contradict the faith. In 1267 and 1268, the Franciscan Bonaventure complained of those philosophers who taught the eternity of the world, the unicity of the intellect, and the idea that the will could be determined by the motions of the celestial bodies.[40] Equally significant, the Radical Aristotelians' effort to separate the conclusions of theology from those of philosophy undermined the Augustinian belief that there was a fundamental unity of truth. Followers of Augustine, such as Aquinas, maintained that when properly practiced, theology and philosophy would necessarily reach the same conclusions and be mutually reinforcing. The Radical Aristotelians' claim that philosophy could only reach necessary conclusions thus undermined the justifications for the practice of natural philosophy being developed within the mendicant orders. Above all, it meant that the latter discipline could not be used in defense of the faith, whether combating Catharism, false philosophical ideas, or the theory and practice of forms of learned magic.

There was little that the opponents of Radical Aristotelianism could do to prevent the arts masters from teaching these ideas in this manner, for they were teaching known errors in the full knowledge that they were wrong. One of the few options open to them was to demonstrate flaws in their account of nature on a purely philosophical basis.[41] This was the tactic assumed by Aquinas when, in 1268, he returned to Paris for a second period as master of theology. Upon resuming his position, he composed no fewer than eleven treatises that sought to clarify specific Aristotelian ideas. At least one, *On the Unicity of the Intellect,* was directed against Siger's teachings. In this work he challenged both the accuracy of Siger's interpretation of Aristotle's thought and its philosophical coherence.[42] The controversy deepened in 1270 when Steven Tempier (1210–79), the bishop of Paris, issued a condemnation of thirteen specific errors taught within the arts faculty, including several drawn from Siger's writings. Tempier also condemned a number of concepts related to the operative arts—for instance, that the human will acts of necessity or that it could be considered passive. The condemnation declared that anyone found teaching or maintaining such propositions would be excommunicated.[43] Tempier's condemnation sanctioned a severe punishment for the act of holding or teaching specific false ideas, but it left unresolved the broader question of how to prevent philosophers from claiming that they could defend arguments that contradicted the faith as philosophically necessary. Moreover, it did not curtail the Radical Aristotelian movement—Siger of Brabant, for one, continued to teach and write.

Philosophy and Magic in the Summa theologica

The issues raised by the Radical Aristotelians informed Thomas Aquinas's thinking as he labored to complete his masterpiece, the *Summa theologica.* He continued working on this text throughout his second period in Paris and indeed after he left to take up his final position in Naples in 1272. Many of the themes that he discussed were directly relevant to the debates in which he had participated at the University of Paris and to the Dominican order's wider mission to promote correct doctrine and root out error. Unlike the *Summa contra gentiles,* however, the *Summa theologica* was not intended as a resource to be used to combat error. It offered instead an introduction to theology, designed to shape the minds of Dominican friars before they pursued their ministry.[44]

One of the key tasks that Aquinas sought to accomplish in the *Summa theologica* was to explain what theology was and how it related to other disciplines. In the opening question, Aquinas defined theology. During this discussion, he refined and elaborated his system of subalternation that governed the relationship between theology and philosophy. Since theology enjoyed a higher degree of certainty than philosophy, it followed that it was impossible to prove any proposition established in the lower discipline that contradicted the truth established in the higher one. Consequently, any philosophical argument that contradicted the truth established in theology should be considered false. Furthermore, any philosophical propositions that contradicted theological truths were not true demonstrations but refutable arguments. Any apparent divergence between the conclusions reached in the two disciplines suggested that the philosopher, not the theologian, was mistaken in his reasoning. Theologians therefore had the right to determine whether the conclusions established by philosophers were true, but not vice versa.[45]

Later in the *Summa theologica*, Aquinas indicated the circumstances in which a false philosophical idea should be considered heretical. He defined heresy as a species of unbelief and further argued that unbelief could manifest itself in one of two ways. The first was the act of willfully refusing to assent to Christ; this was the form of unbelief characteristic of the pagans and the Jews. It was not, however, heresy. The second was when an individual wished to observe the true faith but "[chose] not those things which [were] truly handed down by Christ, but those things which [were] suggested to him by his own mind."[46] For Aquinas, this behavior was "a species of unbelief pertaining to those who profess[ed] the faith of Christ, but corrupt[ed] its dogmas."[47] A heretic was therefore a Christian who willingly chose to accept false belief. Heretics, he added, must not be tolerated. There was a pastoral dimension to this intolerance. Since the heretic's soul was endangered, it was necessary to attempt to reconcile him or her with the faith. Persisting in error was such a grave offense, however, that it deserved punishment, including excommunication and death. For Aquinas, these measures were not only just but also necessary to prevent the individual heretic from corrupting others.[48]

Aquinas refined his definition of heresy in the following article, noting that one could be a heretic either directly—for instance, by making an explicit denial of an article of faith—or indirectly, as in the case of maintaining a view that led to the corruption of an article of faith.[49] It was possible for a philosopher to express a heretical idea in this latter indirect sense if he defended as true a position that either directly contradicted or had implications that ultimately contradicted an article of faith. Aquinas's system of subalternation offered a means for theologians to regulate philosophical arguments about God's creation, including claims about the abilities of humans, angels, and demons to operate therein. Any attempt to present as true an argument that contradicted the faith—one, for example, that denied the existence of angels—would be liable to a charge of heresy. There were limits to the utility of Aquinas's arguments. They could be used only to underpin the prosecution for heresy of those who maintained the absolute truth of philosophical ideas that contradicted the faith but not to discipline the Radical Aristotelians who explicitly stated that their conclusions were false.[50]

Aquinas also discussed the problem of superstition in the *secunda secundae* of the *Summa theologica*. He began by defining superstition as a "vice opposed to religion according to excess" that "offers divine worship either to whom it should not be offered, or in a manner it should not."[51] Invoking the authority of Augustine (*De doctrina Christiana* 2.20), Aquinas identified three species of superstition. The first was idolatry, the act of rendering latria—the reverence that was due only to God—to one of his creatures. The second form of superstition was divination—that is, making explicit or implicit pacts with demons in order to gain privileged knowledge. Finally, Aquinas described observances, which included such acts as the production and use of amulets. Aquinas was in no doubt that superstition led to sin and was a sign of unbelief, but he considered it to be distinct from heresy. During his discussion of idolatry, he explained how the two categories differed. He wrote, "Just as religion is not faith, but a confession of faith by outward signs, so superstition is a confession of unbelief by external worship. Such a confession is signified by the term idolatry, but not by the term heresy, which only means a false opinion. Therefore, heresy is a species of unbelief, but idolatry is a species of superstition." In other words, while heresy was concerned with belief, superstition involved the performance of freely chosen actions that betrayed a lack of true belief.[52]

As part of his discussion of superstition, Aquinas further developed his definition of the legitimate boundaries of the operative arts. One key question that he sought to resolve was whether it was acceptable for Christians to perform any kind of magic that explicitly required the operator to seek the assistance of either angels or demons. Aquinas began by discussing the legitimacy of explicitly necromantic forms of magic. As we have seen, some Christians had claimed that they could command demons to do their bidding. In question 90 of the *secunda secundae*, Aquinas concluded that it was possible for humans to command evil spirits only in order to repulse them. This meant that although the practice of exorcism was legitimate, it was unacceptable for a human to seek to command a demon—for instance, by conjuring one in order to learn or obtain something from it. Such activities, he concluded, invariably led the magician to establish some form of fellowship with the demon, who, in turn, assisted him or her. Since the demons would understand such fellowships as a form of reverence, they were to be considered superstitious.[53]

Aquinas further developed these arguments in question 96 of the *secunda secundae,* which addressed the "superstition of observances." In article 1, he considered the more specific question of "whether the use of the *ars notoria* is not illicit." This was a form of magic that, according to its proponents, worked by contacting angels. Aquinas nevertheless concluded, "The *ars notoria* is illicit and ineffective. It is certainly illicit, because it uses certain things for the purpose of obtaining knowledge, which do not have in themselves the virtue of causing knowledge, such as by the inspection of certain figures, and by the utterance of certain unknown words, and other things of this sort." He reached this conclusion by arguing, in accordance with Augustine, that magical techniques had no inherent efficacy. Instead, the use of words appealed to an intelligent creature, a demon, whose powers made the technique appear effective.[54]

Aquinas also offered a framework for assessing the legitimacy of various avowedly natural operative arts. Whereas Albertus had deferred to the authority of pagan philosophers to decide such issues, Aquinas's views were shaped by his synthesis of knowledge, which allowed him to offer an account of the natural order that was informed by theological precepts. To better understand how this worked in practice, let us consider in detail how Aquinas considered the question: "Whether observances directed to the alteration of bodies, as for the purpose of acquiring health or the like, are unlawful." Throughout the *Summa theologica,* Aquinas used the disputed question method, which was divided into three parts. In the first part, Aquinas proffered various arguments—referred to as objections—for and against the question he was discussing, which did not necessarily reflect his own opinion. In the second, he set out his own views; in the third, he rebutted any objections that he had rejected.[55]

In objection 1, Aquinas put forward for discussion the proposition that observances designed to promote health were lawful, on the ground that "natural things, however, have certain occult virtues, the reasoning of which cannot be assigned by man, just as, for instance, the magnet draws iron, and many other things which Augustine enumerates in *De civitate Dei* 21."[56] In the passage Aquinas cited, Augustine had discussed examples of naturally occurring wonders that lay beyond ordinary human comprehension and wonders achieved through both human artifice and demonic connivance. In objection 2, Aquinas proffered the unattributed opinion that "natural bodies draw certain occult virtues, according to their species, from the influence of the celestial bodies. Therefore, artificial bodies, for instance images, also draw a particular occult virtue from the heavenly bodies for the purpose of causing a particular effect." This opinion appeared to suggest that it was naturally possible for humans to create bodies that could draw celestial influences. This raised the question of the legitimacy of astrological talismans, an issue on which Aquinas had left an ambivalent response in his *Summa contra gentiles*.[57]

To formulate his answer to the question of the legitimacy of observances used to promote health, Aquinas again drew on *De civitate Dei*, book 21. From his reading of Augustine's arguments, he concluded that naturally occurring wonders with occult causes did indeed exist and that it was acceptable for Christians to make use of them. He then turned to consider the legitimacy of observances directed toward improving health. He observed that "with regard to these things which are made in order to induce a certain bodily effect, it must be considered whether they seem to be able to cause these effects naturally. For, if it is so, it will not be illicit, for it is licit to employ natural causes for the purpose of [employing] their particular effects." Aquinas therefore explicitly maintained that it was acceptable for Christians to exploit an object's naturally occurring occult qualities to promote an individual's health.[58]

Aquinas's position left an important question hanging: How would it be possible to distinguish between effects achieved through the manipulation of natural, occult qualities on the one hand and heterodox magical practices on the other? Aquinas's efforts to resolve this question provide an important insight into the methods he used to determine the legitimate boundaries of the operative arts. He stated that if a particular observance undertaken for the purpose of producing a bodily effect "does not seem to be able to cause such effects naturally, it follows that they are not being employed as if for the purpose of causing effects, but only as if they were signs." Citing Augustine's *De civitate Dei*, book 21, he suggested that these signs appealed to a demon who would produce the desired effect on the operator's behalf. In other words, if the causes of a particular effect appeared not to be natural, then they should be considered to be demonic. This observation established an essential set of criteria for examining the legitimacy of operative arts. Any observance of questionable orthodoxy should be investigated to determine whether the effects that their operator claimed they could produce had natural causes. For an art to be defined as explicitly orthodox, its operator needed to prove that its effects had a demonstrably natural cause, even if that cause was itself occult. The corollary of this argument was that if it was not possible to demonstrate that the art's effects were produced by a human's unaided manipulation of nature, then they were likely to have been produced with demonic assistance.[59]

In the final part of the article, Aquinas responded to the objections raised to the original question. In the process, Aquinas offered further guidance on distinguishing between observances that produced effects with natural causes and those with demonic causes. In his reply to objection 1—that there existed natural albeit occult forces—Aquinas again insisted that there was nothing inherently superstitious or unlawful in using naturally occurring things to produce effects. He qualified this statement by adding, "If in truth they are supported by either certain characters, or by certain names, or other various observations whatsoever, which it is clear do not naturally have efficacy, it will be superstitious and illicit."[60] Aquinas's central point was clear: if a human wanted to use natural substances to work effects, he or she should refrain from altering them. Aquinas continued this theme in his reply to objection 2, the suggestion that it was possible to produce artificial bodies that could draw astral influences. He formulated his response by invoking an argument made by Augustine in *De civitate Dei* 10.11, in which the church father had explicitly rejected the idea that human artifice could be used to draw down astral influences for the purposes of altering bodies. On Augustine's authority, Aquinas branded practices such as fashioning talismans—in this instance for the specific purpose of altering health—false and demonic. While Aquinas's views on this matter were ambiguous in the *Summa contra gentiles,* here he left no room for doubt.[61]

In question 95 of the *secunda secundae,* Aquinas applied similar techniques to determine whether it was legitimate for humans to attempt to make knowledge of the future—for example, by practicing arts such as geomancy, haruspicy, and chiromancy. He rejected these arts on the basis that the effects they were supposed to produce (that is, providing privileged knowledge of future contingent events) could not have natural causes. He therefore concluded that any information gleaned from the practice of these arts must ultimately be provided by demons.[62] Although Aquinas had followed Augustine, Isidore of Seville, and the eleventh-century canonists in rejecting almost all forms of divination, he was prepared to concede that it was possible for humans to make some knowledge of the future. He argued that the future could be known in one of two ways: either through knowledge of its causes or in itself. The latter form of knowledge was, he wrote, reserved to God alone, although he might share some of this knowledge with selected humans. This was the gift of prophecy. Aquinas also maintained that by understanding the natural causes of future events, humans could predict their occurrence. Contradicting Augustine's repudiation of astrology, he argued that an astrologer could make two forms of prediction. First, he could foretell with certainty when future events such as eclipses would occur. Second, astrologers could use their knowledge of the stars' movements and their likely effects to make predictions relating to droughts or famines. By these means, physicians too could predict health or death. Aquinas regarded these forms of prediction as "conjectures"—though it was uncertain that the foretold events would occur, it nevertheless remained likely.[63]

In offering this opinion, Aquinas had accepted a far wider range of astrological predictions than the church fathers or the eleventh-century canonists had allowed. Aquinas went still further, however. Although he accepted Augustine's argument that the stars could not be used to make predictions about specific future contingent events and his further assertion that any such knowledge had to be acquired through a demonic pact, he suggested that it was possible to predict some specific future human actions. In an argument that recalled Albertus Magnus's principle of twofold action, he maintained that while the stars could act on the human body and thereby influence the lower faculties of the soul, the will could overrule these inclinations. Consequently, while the stars could incline an individual toward an action, that individual would never lose the ability to resist their influence, meaning that the action would not be necessary. It therefore followed that any prediction an astrologer made concerning future human actions must be uncertain. Nevertheless, since many humans failed to exercise their free will, these predictions were likely to be successful, especially when they were made in relation to the actions of a large population. Although consistent with Albertus's beliefs and those of the author of the *Speculum astronomiae*, these views contradicted the opinion of Augustine, who had denied that the stars could exert any influence over the human body, let alone affect an individual's actions.[64]

The Condemnations of 1277 and Aquinas's Disputed Legacy

As the thirteenth century drew to a close, the use of pagan wisdom, whether philosophy or knowledge of operative arts, remained contested within Christendom. Members of the mendicant orders—including influential theologians such as Albertus Magnus and Thomas Aquinas—had written extensively on these matters. Their views initially had only a limited influence on the society around them. Indeed, the most significant efforts to control the teaching of philosophy and magic were driven by the bishop of Paris, Steven Tempier. In 1277, he issued a further wide-ranging decree condemning certain teachings in the arts faculty of the University of Paris. J. M. M. H. Thijssen has offered a revised reconstruction of these well-known events. He argued that in November 1276 Siger of Brabant and two other arts masters, Bernier of Nivelles and Gosvin de La Chapelle, were summoned to appear before Simon du Val, the inquisitor of France, to answer charges of heresy. Underlining the difficulty of prosecuting those who taught philosophical errors, the men were acquitted. Dissatisfied, Tempier used the documents produced by the trial as the basis for a further investigation into allegations of false teaching in the faculty of arts at the University of Paris. Working with sixteen masters of theology and a number of "important persons," likely to have included the chancellor and the papal legate, Tempier issued a further condemnation of 219 errors in 1277.[65]

Tempier's condemnation opened by chastising those who would discuss various errors contained in works of philosophy, and it continued by criticizing the Radical Aristotelians' suggestion that the arguments presented in philosophical works were so convincing that they could find no answer. Directly alluding to the position adopted by the Radical Aristotelians, the condemnation rejected the argument that it was possible to maintain the view that an opinion was "true according to philosophy but not according to the Catholic faith." Those who made this claim were, according to the authors of the condemnation, in fact dissimulating their true beliefs. The condemnation threatened to excommunicate all who had held, defended, or even listened to such beliefs unless they identified themselves to the chancery of Paris within seven days. It continued by identifying a series of specific heterodox propositions. Some were concerned with cosmological issues (for example, discussion of the eternity of the world). Others implied criticism of the teachings of Aquinas; between fifteen and twenty of the condemned propositions may have been drawn from his work. It is likely that Tempier, following the advice of some theologians and ecclesiastical officials, was seeking to prevent members of the arts faculty from teaching certain Thomist propositions. As Thijssen has argued, this does not necessarily suggest an attempt to mount a retrospective inquiry into Aquinas's thought, but it does indicate that influential individuals harbored reservations about some of his ideas.[66]

Several of the condemnations of 1277 had implications for the theoretical discussion of the boundaries between natural, supernatural, and preternatural causation. Some specifically attacked principles that necessitated the rejection of belief in divine providence. For example, they proscribed teaching the idea that every effect on Earth was governed by the stars and that terrestrial events followed a cyclical pattern that recommenced every thirty-six thousand years when the stars returned to their original positions. By censuring principles of this sort, the authors of the condemnation restricted the arts masters' freedom to teach the most overt forms of Aristotelian naturalism. Other condemnations held implications for the practice of the operative arts, such as astrology. For example, the authors of the condemnation censured the idea that the stars could influence human actions, although their words were moderated by noting, "This is erroneous unless it is understood of natural events and by way of disposition."[67]

The assault on magical practice continued in a further section of the condemnation, which stated, "We likewise condemn the books, scrolls, and leaflets dealing with necromancy, or containing experiments in fortune-telling, invocations of devils, or incantations endangering lives, or in which these and similar things evidently contrary to the orthodox faith and good morals are treated."[68] With these words, the bishop of Paris had condemned a range of divinatory and necromantic practices. The latter category was likely to have included the use of forms of theurgical practice such as the *Ars notoria*, which, irrespective of their operators' claims, were frequently considered works of necromancy by their critics. The condemnation gave no suggestion that the practitioners of these arts had sought to defend them by invoking their distinction between philosophical necessity and absolute truth. In any case, they could only use this defense to justify their discussion of the art's theoretical underpinnings. If they were practicing the techniques described, any defense based on a distinction between philosophical necessity and truth would be unsustainable. This section of the condemnation does, however, underscore the fact that interest in learned magic, whether forms of divination or theurgical practices, had developed in concert with an increasingly sophisticated understanding of pagan philosophy at centers of erudition such as the University of Paris.

Although it is likely that Aquinas would have supported many of the condemnations relating to the operative arts issued in Paris in 1277, his ideas do not appear to have informed the decisions made by Tempier's commission. An inquisitorial handbook dating to between 1260 and 1270, *Summa de officio inquisitionis*, shows that in the later thirteenth century his views also played a marginal role in his fellow Dominicans' pastoral ministry intended to educate and discipline Christian society. This is especially clear in the author's discussion of magic and superstition. Unlike earlier handbooks such as the *Ordo processus narbonensis* or the *Summa de catharis et pauperibus de Lugduno*, this text did include a section dealing with magic and divination, entitled "Form and Manner of Interrogating Diviners and Idolaters." Its author did not, however, offer a general justification for the inquisition of magical activities. Indeed, he did not explicitly refer to Alexander's bull *Quod super nonnullis*, which had permitted inquisitors to investigate forms of magic that manifestly savored of heresy, nor did he argue that these potentially "idolatrous" activities were heretical.[69]

Rather than offering a detailed discussion of superstition and magic, the author of the *Summa de officio inquisitionis* limited himself to proposing subjects that inquisitors could raise with a suspect. His advice appears to have been rooted in the judgments developed in the compendia of canon law compiled during the eleventh and twelfth centuries. Consistent with this tradition, he instructed inquisitors to investigate any acts that involved the worship of demons. He also advised them to raise questions about any practices designed to promote love between men and women or those useful for finding lost treasure or identifying thieves. Enquiries of this nature could have been related to folk or popular magic, but other questions were specifically directed toward forms of learned magic and divination. For instance, the handbook recommended that the inquisitor ask whether the suspect believed in "Egyptian days" or whether they believed that through observation of the movements of the stars it was possible to select propitious days for undertaking a journey or marriage or constructing a building. The author of this handbook did not, however, engage with the detailed reflections on superstition that had begun in the universities. Specifically, he did not discuss the legitimacy of putatively natural operative arts that had occupied Albertus and Aquinas. For example, in his discussion of astrology, he did not consider whether any forms of astral prediction might indeed be natural and therefore acceptable to Christians.[70]

Conclusion

During the thirteenth century, the intellectual culture of the Dominican order developed rapidly, in terms of both its institutions and its ideas. From the 1240s, leading intellectuals in the order, most notably Albertus Magnus and Thomas Aquinas, began to develop new understandings of pagan philosophy and associated operative arts and magical practices, alongside new frameworks for determining their legitimacy. Their systems (especially that of Aquinas) are often presented as definitive statements of the boundaries of orthodoxy. There is, of course, a kernel of truth to this characterization. As we shall see in the following chapter, the ideas of many inquisitors were rooted in Aquinas's thought. Yet this does not mean that Aquinas's ideas were universally accepted as a definitive statement of orthodoxy. Albertus and Aquinas also disagreed about the legitimacy of specific arts. As we shall see in the following chapters, later philosophers and theologians drew on their respective ideas to defend or condemn such practices as the construction of astrological talismans. It was also difficult to determine what a "Thomist" definition of the boundaries of superstition was. Over the course of his lifetime, Aquinas had expressed contradictory views on specific operative arts—for instance, concerning astrological talismans and chrysopoeia. It remained unclear which parts of his corpus should be considered definitive or how to interpret specific ideas. As we shall see in the following chapters, even members of Aquinas's order struggled to determine what should be considered the "true" reading of his work.

✳ 3 ✳

The Inquisition of Learned Magic in the Fourteenth Century

Introduction

The inquisition of magic depended on the ability of officials appointed by the papacy to investigate specific offenses within territories that lay under the authority of local prelates and secular leaders. During the fourteenth century, this authority was contested in two ways. First, both secular monarchs and some members of the episcopate disputed the papacy's ability to grant privileges to the mendicants to undertake their pastoral ministry without episcopal supervision, as well as its authority to allow its inquisitors to operate independently of the local secular and religious powers. These events marked the beginning of a dispute that would continue into the sixteenth century. Second, some Christians challenged papal inquisitors' jurisdiction over magic, which, as we have seen, also derived from papal authority. Responding to this latter challenge in the 1320s, the papacy determined that the practitioners of operative arts that involved making an explicit compact with or an appeal to a demon could be examined in the manner of heretics. The papacy subsequently encouraged secular and ecclesiastical authorities, including inquisitors, to prosecute magic conceived in this manner. Disputes over the inquisitor's authority nevertheless continued through the century, and they provided an important context for Nicolas Eymerich's discussion of magic in the *Directorium inquisitorum*. In this work, Eymerich combined the new papal conception of magic with arguments derived from Aquinas's writings to create an influential justification for the inquisition of magic and divination. As we shall see, however, Eymerich's arguments were based on a selective reading of Aquinas's work, which caused him to articulate criteria for the censorship of magic that differed in significant ways from those set out in the *Summa theologica*.

The Inquisition and Papal Authority

By the late thirteenth century, the papacy had both successfully asserted its supremacy over the Church and established the principle of the freedom of the Church from secular intervention. The popes had also claimed the power to authorize interventions that disrupted existing structures of pastoral care within local dioceses; notably, they invested the friars with the authority to conduct cure of souls and the work of inquisition. The papacy and the mendicants thus developed a mutually supportive relationship. While the friars' activities enabled the papacy to project its power, their privileges depended on papal authority. Consequently, an attack on the mendicants' privileges constituted an attack on papal authority, while constraints on papal authority limited the scope of the mendicants' activities. As we saw in chapter 1, during the thirteenth century a number of secular theologians at the University of Paris, supported by some members of the episcopate, began to question the basis of the mendicants' involvement in the cure of souls at the diocesan level. In response, Pope Martin IV issued the bull *Ad fructus uberes* (1282), which asserted the mendicants' right to preach and hear confession without having first sought permission from the local bishop.[1]

After the accession of Philip IV to the throne in 1285, the French monarchy also began to challenge papal authority. Tensions came to a head in 1294 during a dispute over the monarch's right to tax the clergy during times of war. Boniface VIII was forced to accept the king's demands when Philip prohibited the export of all ecclesiastical tax revenues to Rome, and Boniface's enemies within Italy called for a council that would try the pope for simony and heresy.[2] During this period, Boniface also made further concessions relating to the rights and privileges of the mendicants to offer pastoral care. In 1300, he passed a new bull, *Super cathedram*, which declared that the friars should be free to preach but accepted that they could hear confession and impose penance in a given parish only if they had first been examined by the local bishop. The bull therefore restored some of the power of local bishops while placing clear limits on mendicants' activities within their dioceses and, by extension, on papal authority to intervene in them.[3] These disputes between the French monarchy and the papacy also had direct implications for papal inquisitors, for Philip IV began to attack their privileges in southern France and even sought to have the Dominican inquisitor Foulques of Saint-George dismissed from his post.[4]

As the fourteenth century opened, Philip IV stepped up his attacks on the papacy. He sought not only to further limit papal authority but also to appropriate a portion of it in order to establish what Julien Théry-Astruc has termed a "royal theocracy." Philip IV pursued this strategy through a series of carefully managed conflicts with the papacy, notably placing Bernard Saisset (c. 1232–c. 1314), the bishop of Pamiers, on trial in a secular court and seeking the dissolution of the Knights Templar.[5] Since the Templars, like the mendicants, directly owed their allegiance to the papacy, an attack on them was also an indirect challenge to papal authority. Relations between the papacy and the French monarchy improved in 1305 after Boniface's death and the election of his successor, Clement V (1305–14). Two important developments occurred during his pontificate. First, to successfully broker a compromise with Philip IV, Clement established his court in the city of Avignon. Although Clement had intended this relocation to be only a temporary expedient, the move became permanent under his successor, John XXII.[6] Second, in 1308, Clement V agreed to hold provincial councils to try individual Templars and to convoke a universal council to offer a final judgment on the order. During the Council of Vienne (1311), Clement ordered the dissolution of the Templars, a move that was resisted by the assembled fathers.[7]

While the Council of Vienne was primarily concerned with the examination of the Templars, it also dealt with a series of other matters. Perhaps reflecting the diminished power of the papacy in relation to the French monarchy, several decrees introduced measures that appeared to moderate aspects of papal power. Decree 10, for example, reiterated the terms of *Super cathedram,* the earlier bull limiting mendicant cure of souls.[8] In decree 26, the council also implemented measures to regulate the practice of inquisition. This was a significant development, for previous universal councils had not concerned themselves with such matters. The decree opened with the observation that "the apostolic see has received many complaints that some inquisitors, appointed by it to suppress heresy, have overstepped the limits of the power given to them." The decree did not indicate the limits of inquisitors' authority but noted that on occasion inquisitors "so enlarge[d] their authority that what ha[d] been wisely provided by the apostolic see for the growth of the faith, oppresse[d] the innocent" and ultimately damaged the faith. It is perhaps no coincidence that in his earlier attempt to restrict the power of inquisitors within his kingdom, Philip IV had leveled similar complaints. To prevent inquisitors from acting in this manner in future, the decree proposed that the episcopate should play a larger role in the practice of inquisition. It stated that although bishops and inquisitors could each act independently to conduct investigations, neither had the power to imprison or torture suspects without having first consulted the other.[9]

A number of the decrees of the Council of Vienne may have sought to encourage closer supervision of papal institutions by local ordinaries, but others also demonstrated continuities with the earlier papal reforming agenda. Some decrees promoted monastic reform, for example. Others sought to extend the process of reform to the laity by expanding the programs of pastoral reform and the associated techniques of social discipline established at the previous two Lateran Councils. Several specific decrees were intended to prevent the circulation of heretical ideas and practices. Decree 28, for instance, sought to control the Beguine movement that had begun in the Low Countries in the twelfth century, which encouraged groups of the laity to live in semimonastic communities. Further decrees were intended to curb sinful practices. Decree 29, for example, condemned the practice of usury and branded as heretics those who pertinaciously maintained that it was not a sin. Other decrees expressed a desire to spread the faith to nonbelievers such as Jews and Muslims. Decree 24 promoted the establishment of schools to teach Hebrew, Arabic, and Chaldaic in the university towns of Paris, Oxford, Salamanca, and Bologna in order to support efforts to proselytize among non-Christians. Despite these initiatives, it appears that during the fourteenth century, the friars' preaching was primarily directed toward Christians. Indeed, their actions against the Jewish community were largely limited to efforts to control the circulation of the Talmud and root out Judaizers. There was no concerted effort to win converts.[10]

The Inquisition of Learned Magic in the Early Fourteenth Century

The Council of Vienne had passed a series of decrees designed to promote social and spiritual discipline in Christendom, but they had not referred to the problem of superstitious magic. Purging society of these ideas and practices nevertheless became increasingly important to the papacy during the fourteenth century. These ambitions are reflected in the papacy's efforts to assert inquisitors' authority to investigate cases of superstition. Alexander IV's bull *Quod super nonnullis* (1258) had permitted inquisitors to investigate cases of magic that "manifestly savored of heresy," but it did not define this ambiguous term. The legal basis of the inquisition of magic was clarified when Pope Boniface VIII authorized the production of a second volume of papal decretals, the *Sextum decretalium liber* or *Liber sextus*. The new volume included the *Quod super nonnullis*, with an accompanying gloss that expanded on the bull's original definition of heretical magic and determined the extent of an inquisitor's authority to prosecute such cases. It declared that to savor of heresy "is, for example, to utter prayers near abominable altars of idols, to offer sacrifices, to consult demons, to receive their responses. Or they associate themselves with heretics in order to cast lots, or they make predictions with the body and blood of Christ, or when casting lots, so that they might have a response, they re-baptize boys or something similar."[11]

The gloss on *Quod super nonnullis* offered important clarification on aspects of the inquisitor's authority to investigate magic, but it left a number of issues unresolved. When the gloss defined the inquisitor's authority to prosecute cases of heretical magic, it emphasized the importance of investigating practices that included explicit ritual elements, whether praying at altars or rebaptizing children. Such activities required investigation because they were likely to involve the suspect making an explicit pact with a demon. Framed in this manner, the gloss suggested that the inquisitor had no mandate to investigate magical practice per se but only those cases of magic in which the operator was suspected to have consciously invoked demons or to have misappropriated Christian ritual. Importantly, the gloss made no reference to issues raised about contested operative arts by theologians such as Albertus Magnus and Thomas Aquinas during the previous century. Although the gloss did not explicitly make this point, its wording implied that operative arts that did not explicitly invoke demons—such as astrology, physiognomy, or geomancy—did not manifestly savor of heresy, which in turn suggested that they were not subject to inquisitorial jurisdiction.

Despite these legal innovations, trials for sorcery and magic remained rare during the first third of the fourteenth century. Having completed a Calendar of Witch Trials, Richard Kieckhefer concluded that between 1300 and 1330, there was on average only one trial per year in all courts across western Europe. He also noted that rather than being driven by a desire to eradicate magic from Christian society, most of these cases had a political character. Often, the victims (and sometimes the suspects) were members of the ecclesiastical and secular elite. To take two well-known examples, during their dispute Philip IV leveled charges of sorcery against Boniface VIII and subsequently accused the Templars of diabolism. Kieckhefer did not, however, find any convincing evidence to suggest that any trials for sorcery were heard in inquisitorial tribunals.[12] The reasons behind inquisitors' reluctance to prosecute cases of magic among the wider population of Europe remain unclear, but it may well reflect lingering questions about their authority to investigate these matters and the correct definition of illegitimate magic.

To take an important example, the status of several forms of ostensibly natural healing magic remained unclear. This is evident in the reaction to the writings of Arnald of Villanova (1240–1311). Having studied at the University of Montpellier, he subsequently played a leading role in developing medicine at this institution. He distinguished himself in medical practice, serving for a time as physician to the court of Aragon and treating popes Boniface VIII (1294–1303) and Benedict XI (1303–4).[13] Arnald also authored a number of medical texts, some of which contained descriptions of magic. In his most famous work, *De regimine sanitatis*, a general handbook for preserving health, he remarked on the occult power of certain stones. For example, in a discussion of the factors peculiar to the health of women, he noted that placing the stones serapinum and hypericon in their room would force any demons to flee. Arnald did not offer any explanation for how these stones might produce this effect, nor did he offer any theological justification for this belief. It is possible that he did not feel the need to justify these practices, for as we have seen, according to eleventh- and twelfth-century compendia of canon law, such practices were acceptable.[14] Elsewhere, Arnald considered the power of signs. In one work, *Antidotarium*, he argued that certain signs, whether naturally occurring or produced by human artifice, could derive from the stars the power to neutralize the effects of an ingested poison.[15] Arnald was a controversial figure in his own lifetime and was investigated by ecclesiastical authorities. Significantly, however, these authorities were concerned by his views on issues such as eschatology and Church reform rather than the claims made in his philosophical and medical works.

Intellectuals could, it seems, continue to read about and discuss ostensibly natural operative arts including talismans within early fourteenth-century Europe, without arousing the suspicion of theological authorities such as the Inquisition. This situation may be partly explained by the fact that many contemporaries considered these arts to be natural and therefore legitimate. Frank Klaassen has shown that manuscripts relating to the production of astrological talismans were widely copied in medieval Europe and that they were treated as part of discourses on *naturalia*.[16] The problem remained, however, that important theologians such as Albertus and Aquinas held conflicting opinions about whether astrological talismans could create their effects naturally. Defending the use of talismans therefore required not only philosophical argument but also the selective deployment of theological authority. In texts such as the *Conciliator* and the *Lucidator*, the physician and philosopher Pietro d'Abano (1257–1316) drew extensively on the *Speculum astronomiae*, a text ascribed to Albertus, to discuss the medical uses of astrology and the means to construct astrological talismans for the purposes of improving health. Echoing Albertus's opinions, he maintained that these arts functioned by natural means and that for this reason it was acceptable for Christians to use them. Through these and other works, Pietro contributed to the construction of a highly sophisticated complex of astrological-medical ideas that became embedded in the curricula of Italian universities such as Padua and Bologna.[17]

John XXII and the Definition of Heretical Magic

The papacy's campaign against magic grew more intense during the pontificate of John XXII, who was elected pope in 1316. Just a year after his accession to the papal throne, evidence came to light of a plot to kill the new pope and several of his confidants. It was led by Hugues Géraud, bishop of Cahors, who sought to exact revenge for papal threats to depose him from his office for the offense of simony. To carry out his scheme, the bishop had acquired not only arsenic but also wax dolls intended to be used to work maleficiant magic against the pope. The conspirators sought to baptize these figures with the name of the intended victim, thereby allowing the devil to recognize and kill the individuals that they represented. Having successfully tested their techniques on the pope's cardinal-nephew, they resolved to implement their plan. The conspirators smuggled the effigies into Avignon, but the plot was discovered before they could strike. During the subsequent investigation, Géraud confessed to his crimes. He was burned alive in September 1317.[18]

Following the conclusion of this trial, John began to encourage both bishops and inquisitors to root out any other would-be malefactors. In February 1318, he sent letters to the bishop of Fréjus pressing him to prosecute cases of necromancy and the following July to the bishop of Pamiers, Jacques Fournier (1285–1342), a Cistercian and later Pope Benedict XII, urging him to proceed against three individuals whom he accused of offenses including creating images and making incantations.[19] In 1320, William, cardinal of Santa Sabina, wrote to Jean de Beaune (the inquisitor of Carcassonne) and Bernard Gui (1261/2–1331, the inquisitor of Toulouse), informing them of the pope's desire that the "malefactors, those who infect the Lord's flock, flee from the midst of the House of God."[20] Santa Sabina's use of the term *malefactors*—*maleficos*—gives a clear sense of John's concerns: he specifically wanted to remove from Christian society those who sought to perform harmful magic.[21] Despite this specific emphasis on *maleficia*, the conception of magic conveyed in Santa Sabina's letters was in other respects consistent with the one sketched in the 1298 gloss on *Quod super nonnullis*. He urged the inquisitors to take action against those who made explicit pacts with demons, baptized an image or had one baptized on their behalf, repeated sacraments such as baptism or confirmation, or used consecrated hosts or other paraphernalia of Christian ritual to perform works of sortilege or *maleficia*. In other words, the letter encouraged only the prosecution of offenses that explicitly involved the invocation of demons or the subversion of Christian ritual. The letter did not suggest that inquisitors or bishops should investigate operative arts such as alchemy, astrology, or even forms of putatively natural image magic.[22]

John XXII's efforts to encourage inquisitors and bishops to investigate maleficiant magic appears to have had, at most, a limited effect. Even in areas under the jurisdiction of bishops and inquisitors who had received letters from John XXII and Santa Sabina, the number of trials for magic did not increase. The extant records suggest that Bishop Fournier never executed anyone for magical offenses. During his term as inquisitor of Toulouse (1305–25), Bernard Gui prosecuted 603 cases; none were for magic.[23] Perhaps frustrated by the fact that the inquisition of magic had not gathered pace, in the autumn of 1320, John XXII convened a theological commission charged with the task of determining whether four specific types of magic should be classified as heresies. These were baptizing images, rebaptizing Christians, using the consecrated host to cast spells, and invoking demons. The commission was composed almost entirely of members of the regular clergy, and its members included Fournier.[24] As we have seen, a number of the activities that John XXII asked his commission to consider had been referred to as examples of heretical magical practice in the gloss on *Quod super nonnullis,* contained in the *Liber sextus*. The fact that John XXII created a commission to consider these matters again suggests that the question of whether these activities should be considered heretical—and therefore within the inquisitor's purview—remained contested. It would indicate that the gloss on *Quod super nonnullis* was not considered a definitive guide to forms of magical practice that manifestly savored of heresy.

The members of the commission all accepted the proposition that the activities that they had been asked to consider could be defined as heresies, but they each supported it with varying degrees of enthusiasm. Drawing on the opinions provided by his commission, John drew up the bull *Super illius specula,* which he promulgated in 1326. Once again, a number of the ideas that it set forth were consistent with those elaborated in the gloss on *Quod super nonnullis*. Notably, it specifically proscribed offering sacrifices to demons or worshipping them. It offered more detail, however, by providing specific examples of illicit magic. They included producing images, having someone produce images on one's behalf, and binding demons into rings or phials for the purpose of acquiring information or assistance. The bull also warned that individuals should refrain from teaching, discussing, and, above all, practicing any of these arts. The bull left ambiguous the precise status of those who performed magic. Although it stated that anyone who did not abide by this judgment would receive "each and every punishment . . . that heretics deserve by law," it did not specifically argue that anyone who practiced these arts was indeed a heretic.[25]

Super illius specula built on the principles established by *Quod super nonnullis*, but it shared many of its limitations. Although the bull was specifically designed to address the forms of magic practiced by learned elites, it did not amount to a comprehensive assessment of learned magic. The bull prohibited a series of practices associated with ritual magic, but it did not elaborate or endorse any particular criteria for determining the legitimacy of contested operative arts such as astrology, physiognomy, or geomancy. Furthermore, it did not specify the grounds on which the particular forms of magic that it named were to be condemned. To take one specific example, the bull condemned the production of magical images, but it was unclear from the text whether this prohibition referred only to explicitly necromantic images or whether it also included ostensibly natural images. As we have seen, while Albertus and later scholars such as Pietro d'Abano and Arnald of Villanova believed that it was legitimate to make astrological images, Aquinas maintained that they produced results only because the operator had made a tacit pact with a demon. Contemporary trial testimony relating to the production and use of images suggests that, in practice, Aquinas's concept of the "tacit" pact was rarely if ever invoked to condemn these arts. Inquisitors tended to prosecute only those cases in which, in addition to discovering images, they found further evidence that the operator was also performing rituals designed explicitly to invoke demons.[26]

Bernard Gui and the Practica inquisitionis heretice pravitatis

John XXII's letters exhorting inquisitors and bishops to investigate *malefici* do not appear to have caused an increase in investigations. The evidence from inquisitorial manuals suggests, however, that some of the ideas that John XXII advocated were becoming embedded in the theoretical discussion of the inquisition of heresy. This is evident in one of the most influential inquisitorial manuals of the age, the *Practica inquisitionis heretice pravitatis*, written by the Dominican Bernard Gui, inquisitor of Toulouse. Gui worked on the text of this book intermittently throughout his tenure as inquisitor, completing it at some point between 1323 and 1324—that is, after he had received Santa Sabina's letter concerning maleficiant magic but before the promulgation of *Super illius specula*. The manual was divided into five parts. The first two contained a series of formulas for drawing up inquisitorial documents, largely based on Gui's own records and sentences. The third described the conduct of sermones generales, the ceremonies at which repentant heretics received their sentences. The fourth part provided a discussion of the office of the inquisitor, which was largely derived from an earlier Italian work, *De auctoritate et forma inquisitionis.*[27]

The final part of Gui's text was a treatise that categorized the various forms of deviance that an inquisitor might be expected to face in the course of his duties and sketched in relief the society that they might help create. Its opening sections described the fundamental beliefs of the Cathars, the Waldensians, the Beguines, and the Dolcinites. A further section described the inquisition of Judaizers—that is, relapsed Jewish converts to Christianity—and the investigation of Jewish literature such as the Talmud. Eymerich's discussion of the Jews indicates how the mendicants were developing their own distinct reforming agenda as early as the fourteenth century. On the one hand, the inquisition of Judaizers was consistent with papal ambitions. Although the papacy continued to promote official tolerance of Jews within Christian society, it supported attempts to convert them and prevent their reversion to their original faith. Pursuing this latter aim, in 1267, the papacy had promulgated the bull *Turbato corde,* giving inquisitors the authority to investigate Judaizers. The decision to include material on the Talmud, on the other hand, illustrated the sometimes contrasting objectives of the papacy and the mendicants. Since the early thirteenth century, the friars had encouraged the investigation of Talmudic literature, which they believed both insulted and endangered the Christian faith. Initially, the papacy had supported these measures, but by as early as the 1240s direct papal intervention in efforts to restrict the Talmud became rare. Despite the papacy's ambivalence, during the fourteenth century, southern French inquisitors (notably Gui) continued to play a leading role in efforts to gather up and destroy the Talmud and to a lesser extent prosecute Judaizers.[28]

In the shortest section of Gui's treatise on heresy, he discussed "Sortilegers, Diviners and those who summon demons." The inclusion of this section was perhaps surprising considering how little energy Gui appeared to have expended on investigating cases of magic. Indeed, the fact that he even chose to consider magic and divination may reflect John XXII's influence. The evidence from Gui's manual nevertheless suggests that inquisitors' thinking about magic had not developed significantly since the thirteenth century. There is certainly no indication that Gui had engaged with the sophisticated theoretical discussion of magic undertaken by theologians such as Albertus Magnus or Thomas Aquinas. This may be explained by the fact that although the two friars, and especially Aquinas, were highly revered within their orders by the mid-fourteenth century, their discussions of magic were not central to Dominican education. In the field of theology, the central teaching text remained Peter Lombard's *Libri quattuor sententiarum*. Aquinas's *Summa theologica*, which contained his most sustained engagement with magic and divination, was read and taught, but it had not yet attained the central importance in Dominican education that it would later achieve.[29]

The discussion of magic contained in Gui's handbook was, like earlier works such as the *Summa de officio inquisitionis*, restricted to a series of questions that an inquisitor should raise with those suspected of sorcery, fortune-telling, or invocation. He instructed that suspects should first be asked what they knew of these arts, how many they knew, and from whom they had learned them.[30] Gui continued by providing a series of more specific questions that inquisitors might raise during their investigations. He suggested asking suspects what they knew of matters including placing children under or releasing them from a spell; impregnating the sterile; enchantments and conjurations using songs, fruits, or plants; and curing disease by conjuration. Most of his questions therefore appeared to relate to forms of folk magic, and they included techniques intended to produce not only maleficiant but also beneficent ends. Following the precedents established in compendia of canon law, Gui was primarily concerned with how magical effects were produced rather than their intended purpose. Gui did not specifically state why he was concerned by these particular practices, but it is likely that either he or his fellow inquisitors had encountered them during their work.[31]

Many of the ideas that Gui initially expressed in his account of magic were broadly consistent with those set down in the earlier inquisitorial manual *Summa de officio inquisitionis*, discussed above. As Gui's discussion developed, it began to demonstrate the influence of the new papal concept of magic. The accused, he wrote, "should be examined most particularly about these things that savor of superstition in any possible way, either irreverence, or injury concerning the sacraments of the Church and above all regarding the sacrament of the body of Christ, and also divine worship and holy places."[32] Gui's reference to investigating practices that "savor of superstition" recalled (although it did not fully match) the language of Alexander IV's bull *Quod super nonnullis*, which had referred to offenses that savored of heresy. He also instructed the inquisitor to inquire whether the accused possessed consecrated hosts, or holy oil, questions intended to expose evidence of the subversion of Christian ritual. In this respect, his proposed questions reiterated at least one of the concerns expressed in the gloss on *Quod super nonnullis* and Cardinal Santa Sabina's letter of

1320. Furthermore, echoing John XXII's previously stated concerns, Gui instructed that the inquisitor should question the accused about practices such as the baptism of images made of wax or any other material, the manner in which those images had been baptized, and how they were intended to be used. Gui also urged inquisitors to ask about lead images, their means of production, and their intended purpose.[33]

In his manual, Gui also provided a number of *modi abjurandi*—that is, formulas for abjuring specific offenses. They included a model text to be used for abjuring sortilege, divination, or invocation. He explained its purpose to the reader in the following manner:

> Means of abjuring the pestilence and error of sortilegers or diviners and invokers of demons, especially where it savors of heresy against the truth and piety of the sacraments of the Eucharist or baptism or any of the other sacraments, or where in the invocation of demons sacrifice or offering was shown or made to a demon or any other error clearly containing [something] contrary to the faith.[34]

Echoing the new papal definition of heretical magic, Gui indicated that the abjuration should be used to recall to orthodoxy those Christians who had engaged in forms of magic that involved either the invocation of demons or the perversion of ritual. This was also broadly consistent with his earlier description of magic, albeit with a subtle difference. He had previously described such activities as savoring of "superstition"; here, however, he explicitly characterized them as savoring of "heresy." Although it was clear that Gui believed inquisitors could investigate these arts as heresies, his precise understanding of the terms *superstitio* and *haeresis* remains unclear from the text.

We gain further insight into Gui's understanding of magic from the text of the abjuration. It demanded that the penitent should swear, "I abjure generally all condemned sortilege, especially that which is arranged for the purpose of procuring any illicit or harmful effect."[35] The injunction that the penitent should renounce "condemned sortilege" could be understood to encompass magic that involved invocation, while the reference to procuring illicit effects certainly reflected the spirit of John XXII's letters and his subsequent bull. Nevertheless, the abjuration did not specify precisely which practices were to be considered forms of "sortilege," nor did it indicate the basis on which they were to be condemned. It made no specific reference to ostensibly natural arts such as astrology, geomancy, or physiognomy, which had been recovered over the preceding centuries, or indeed to other arts such as alchemy. Gui's model abjuration did, however, require the penitent to abjure "the art and means of making images of lead, or wax, or whatever other material for the purpose of procuring whatever illicit effects." This formulation was nonetheless ambiguous. It could have referred either to the practice of making figures for baptism or to the construction of necromantic images that John XXII had condemned. Equally, it could have referred to the production of putatively natural images designed to produce effects. In this model abjuration, Gui left the question of the validity of this latter class of images unresolved.[36]

The Pursuit of Magic, 1330–75

Despite the papacy's efforts to encourage them, trials for magic in western Europe appear to have become less common between 1330 and 1375. Richard Kieckhefer has argued that this overall decline was caused by the end of politically inspired accusations of sorcery and maleficia that had characterized the prosecutions of the first third of the century. Within Italy, there may have been as few as four trials during this period. They included two inquisitorial processes, one in 1360 and the second in 1370, which were possibly the only examples in all of Europe during this period. While the extant evidence suggests that the prosecution of magic in late fourteenth-century Europe was rare, it remains possible that the prosecution of magic was more common than we realize. Many relevant documents have been lost or destroyed—for example, the records of Tuscan inquisitorial and episcopal courts for the fourteenth century no longer survive. However, contextual evidence suggests that these documents would not substantially alter the existing picture. Secular authorities, at least, seem to have prosecuted magic rarely. Using the largely complete archives of the communal courts of Florence, Gene A. Bruckner found that the magistrates heard no cases of magic between 1343 and 1375. A similar picture emerges in Perugia, where the 1347 trial of Riccola di Puccio di Pisa is the only extant record of a sorcery trial heard by the civil courts.[37]

Although unusual, the trials for sorcery and magic that did take place in this period provide important evidence to suggest that ecclesiastical concepts of illicit magic were beginning to enter civic courts. This was evident in the 1347 trial of Riccola. The records reveal that certain residents of Perugia had approached Riccola to work magic on their behalf, often to make one individual either fall in love with or come to hate another. As we saw in chapter 1, civic law had been primarily concerned with maleficiant magic—that is, magic designed to inflict harm—and had not tended to be interested in such acts as casting love magic. The documents recast Riccola's acts by frequently stating that she had worked magic "in order to harm and for the purpose of harming" (*ad nocendum et causa nocendi*). The documents also record that the investigators found that she fulfilled these requests by making magical objects and powders, a process that involved reciting incantations designed to invoke demonic agency. The activities that Riccola was alleged to have performed—working maleficia by demonic means—were therefore consistent with the practices identified as a cause for concern by the papal court and recorded in inquisitorial manuals.[38]

Analysis of one of the two inquisitorial trials for magic that occurred in Italy, that of Benvenuta Benincasa conducted in Modena in 1370, shows that this new papal conception of heretical sorcery was also gradually being assimilated into the practices of papal inquisitors. The inquisitors interrogated Benvenuta about her use of various forms of magic. They included conjuring spirits to discover who had inflicted *maleficia* on a third party, with the express intention of relieving their suffering. In one important passage of the trial, the inquisitor asked her "how she can believe that she does meritorious things by invoking malign spirits, which is something that is prohibited by the Church and condemned as heresy." The case therefore turned on the question of whether it was legitimate for Benvenuta to invoke spirits to achieve benevolent ends. Irrespective of Benvenuta's good intentions, the inquisitor's question revealed that they were concerned by the act of communicating with spirits rather than the intended outcome. The inquisitors found her guilty, and she subsequently abjured her offenses.[39]

Benvenuta's case illustrates the fact that by the 1370s at least some inquisitors had accepted the principle that the operators of magic could indeed be investigated and prosecuted in the manner of heretics. Yet while the papacy might have asserted that it was the inquisitor's right, if not his duty, to investigate and prosecute cases of magic, it appears that some continued to resist this idea. In 1374, Pope Gregory XI (1370–78) wrote to the Dominican inquisitor of France, Jacque de Morey, stressing the inquisitor's obligation to investigate anyone suspected of invoking demons. He also lamented that "some people, and also sometimes the learned, oppose themselves to these matters, asserting that these canonic decrees do not pertain to your office according to canon law." Gregory's letter reveals that unidentified individuals, including members of the learned elite, remained unconvinced by the jurisdictional claims advanced by the papacy. This resistance to the inquisition of magic may explain why prosecutions for magic remained relatively low.[40]

Nicholas Eymerich and the Definition of Heretical Magic

The ongoing disputes about the inquisitor's jurisdiction revealed by Gregory XI's letter to Morey provide an important context for the discussion of magic in the influential handbook the *Directorium inquisitorum*, written by Nicholas Eymerich (1316–99). Since 1358, Eymerich had served as the inquisitor general of Aragon, a role that he executed with marked enthusiasm. While his zealous prosecution of heretics earned him the plaudits of the papacy, his willingness to investigate his fellow clerics created powerful enemies. His efforts to investigate the teachings of the Franciscan Nicholas of Calabria in the late 1350s brought him into conflict with King Peter IV of Aragon. These tensions once more came to the fore when he sought the condemnation of the work of another revered Franciscan, Ramon Lull (c. 1232–c. 1315). Possibly encouraged by Eymerich, Pope Gregory XI ordered the investigation of Lull's works. Partly in response to these events and partly for his role in pursuing individuals closely linked to the crown, Peter IV expelled Eymerich from Aragon. The latter fled to Avignon, where he composed his *Directorium* in 1376.[41]

Like earlier examples of the genre, Eymerich's handbook contained a wealth of information dealing with the definition of heresy and procedures for conducting trials. His preoccupations reflected aspects of the Dominicans' reforming agenda. He asserted the right of inquisitors to investigate non-Christians, including Jews and Muslims, by setting out the basis for prosecuting lapsed converts to Christianity and extending inquisitorial power to encompass acts of blasphemy against the Christian Church.[42] Eymerich also demonstrated his commitment to purging Christian society of superstitious magic. His discussion of these matters is contained primarily in the second book of his text, specifically in question 42, *De Sortilegis & divinatoribus* ("On Sortilegers and Diviners"), and question 43, *De invocantibus Daemones* ("On the invokers of demons"). Although Eymerich did not devote much space to the consideration of magic, he offered a far more detailed account than had been previously attempted either by Gui or in any earlier papal document. Importantly, his discussion also indicated the beginning of attempts to incorporate ideas formulated by Albertus Magnus and Thomas Aquinas concerning the boundaries of magic into the practice of inquisition.[43]

In both questions 42 and 43, Eymerich considered the inquisitor's authority to prosecute magic and divination, but his discussion was far more sustained and detailed in question 43. Here, his account came close to a treatise, one that offered detailed distinctions buttressed by a range of theological and legal authorities. Eymerich began by identifying the different sources from which he had gathered his personal knowledge of invocation. His first set of sources was textual. He noted that he had read two manuscripts on magic: first, the *Clavicula salomonis*, a work of ritual magic that had come to notoriety by the late thirteenth century, and second, *Thesaurus necromantie a necromanticis* (probably the *Thesaurus spirituum*). The knowledge he derived from reading these works was, he noted, complemented by a second source of information: the experiences that he had acquired as a practicing inquisitor. His discussion was informed both by cases of invocation that he had investigated personally and by cases about which he had learned from his fellow inquisitors. Although Eymerich did not explicitly acknowledge this point, his conception of heretical magic was also strongly influenced by the complex of ideas elaborated by the papacy since the late thirteenth century.[44]

Drawing on these sources, Eymerich identified specific forms of invocation. The first involved humans offering latria to demons—that is, the form of reverence that was due to God alone. Examples of this practice included instances when magicians offered sacrifices or adoration to demons or made "detestable orations" to them. The second was when humans offered dulia to demons—that is, forms of reverence due to the saints or the Virgin Mary. Eymerich noted that these and other similar execrable practices were to be found in works such as the *Clavicula salomonis*. In each of these examples, Eymerich maintained the established principle that inquisitors should prosecute cases of magic that involved the invocation of demons, but his discussion refined the idea. For Eymerich, the defining feature of a heretical practice was not engaging in rituals per se—however dubious they might appear—but the conscious act of offering either dulia or latria to demons.[45]

Eymerich proceeded to explain why cases of invocation that involved offering latria and dulia to demons should be regarded as a heresy and hence why they were subject to inquisitorial jurisdiction. In the case of latria, his discussion opened by invoking the authority of Augustine, specifically book 10 of *De civitate Dei*. The next text that he cited was Aquinas's commentary on *Isaiah*. The inclusion of Aquinas was in itself significant. While earlier handbooks, such as Gui's, had not referenced Aquinas's works, Eymerich's work was notable for the frequency with which he invoked his illustrious forebear's authority. In this particular passage, he continued, Aquinas had maintained that it was always a sin to ask demons for any information and that such practices constituted an apostasy from the faith. Underscoring the continuity between his sources' respective positions, he noted, "As Augustine says, here too says blessed Thomas" (*ut dicit Augustinus haec beatus Thomas*). He went on to cite Aquinas's commentary on the *Sentences* and works by other theological authorities such as Albertus Magnus and Bonaventure. He then proceeded to present the opinions of a series of legal authorities. This section culminated with the presentation of the bull *Super illius specula*, which he reproduced in full.[46]

Setting out these authorities in this manner allowed Eymerich to construct a compelling genealogy of Christian theological and legal thought pertaining to magic. He demonstrated the existence of a clear and consistent approach that could be traced from the patristic era through the work of authoritative contemporary theologians such as Aquinas and Albertus, up to the point that it was inscribed in canon law by the papal bull *Super illius specula*. While persuasive, Eymerich's account was in one important respect misleading. He was certainly correct to assert that authorities such as Aquinas and Augustine had explicitly rejected the idea that Christians could legitimately have any form of contact with demons. They had, however, argued that performing these arts constituted superstitious rather than heretical behavior. Eymerich did not draw attention to this discrepancy, nor did he specifically address the distinction between superstition and heresy. By presenting Augustine's and Aquinas's views in this manner, Eymerich suggested that, like him, they had considered acts of invocation to be heretical. Irrespective of whether Eymerich deliberately chose to reconstruct Augustine's and Aquinas's ideas in a misleading way, his account invoked their authority to support his particular, chosen definition of magic and divination. Perhaps more importantly, Eymerich had also been able to suggest that the relatively novel definition of magic contained in *Super illius specula* was in fact entirely consistent with, and indeed constituted the culmination of, Church tradition and canon law. Whether consciously or not, he had justified the inquisition of magic by, in effect, effacing earlier carefully drawn distinctions.[47]

Eymerich's primary concern in question 43 was legitimating the inquisition of forms of ritual magic, and he did not discuss the orthodoxy of operative arts with disputed causes. He did, however, engage with these issues to a limited extent in the preceding question, which was concerned with divination. His discussion began, like his account of invocation, by addressing the question of whether sortilegers and diviners were to be considered heretics. To formulate his answer, he distinguished between different types of sortilegers and diviners, differentiating between those he categorized as "pure" and those he termed "heretical." He gave two examples of practices that fell within the former category. The first were "those who act purely from the art of chiromancy, who from the lineaments of the hand make divinations and judgements about natural effects and the condition of man."[48] The second kind of "pure" operators were those who sought to gain knowledge by drawing lots. Eymerich concluded that inquisitors need not concern themselves with either of these classes of activity, on the basis that they did not involve an overt ritual element. The inquisitor's primary concern, he argued, should be to investigate only those divinatory arts that included explicit forms of invocation. The example that he gave was the practice of rebaptizing babies to divine knowledge of future things or to understand private thoughts. Such acts of divination or sortilege, Eymerich concluded (echoing the language of Alexander IV's bull *Quod super nonnullis*), "indeed manifestly savor of heresy."[49]

These principles informed the advice that Eymerich imparted to inquisitors investigating magic. They should, he instructed, examine the evidence relating to each individual case—whether this derived from a suspect's confession, legitimate witness testimony, or factual evidence—in order to search for proof that the accused had directly engaged in either dulia or latria. He urged inquisitors to investigate whether the suspect had practiced forms of divination with explicitly necromantic elements and whether the accused had practiced sortilege or divination "into which they have mixed in some way, things that manifestly savor of heresy."[50] Eymerich therefore maintained that inquisitors should investigate even those who had practiced ostensibly natural forms of divination in order to see whether they had added an extra ritual element. Those found guilty of this crime were to be considered heretics and given the choice of abjuring or being consigned to the secular courts for punishment. These punishments, Eymerich added, were justified according to the provisions of the papal bull *Ab abolendam*, which had established the use of inquisition in clerical courts, although not on account of the act of divination itself but because of the additional heretical practices. He continued that in those cases where it was not possible to prove that the diviner had acted in a heretical manner but there were nonetheless clear indications that this was the case, the individual in question was to be considered vehemently suspected of heresy. They too would be obliged to abjure their crimes. Finally, in cases where there were no clear indications that the suspect had engaged in heretical activity but there were rumors of such an offense, the suspect would be subject to the *purgatione canonica*.[51]

During the discussion contained in question 42, Eymerich did not cite Aquinas's works directly, but his argument recalled them in one fundamental respect. By acknowledging the existence of "pure" forms of divination—that is, forms of divination uncontaminated by demonic assistance—Eymerich appeared to echo the Thomist-Augustinian position that it was acceptable for humans to make use of naturally occurring effects in order to work wonders. As we have seen, Aquinas had accepted the legitimacy of some predictive arts, arguing that Christians could use an understanding of the natural causes of effects to predict their future occurrence. This reasoning allowed him to legitimate some arts designed to foretell the future, such as limited forms of predictive astrology. The arguments put forward in Eymerich's account of divination had clear affinities with those developed by Aquinas, but there were also points on which they disagreed. First, Aquinas did not talk of legitimate forms of "divination." He used this term to describe only those forms of prediction that he condemned, not those that he considered natural and therefore legitimate. The two Dominicans also held contrasting opinions of the legitimacy of specific arts, notably chiromancy. As we have seen, although Eymerich accepted its limited use, Aquinas had simply considered it to be an illegitimate art designed to make privileged knowledge of the future.[52]

There also remained significant differences between the methods that Aquinas and Eymerich chose to determine the legitimacy of a given operative art. Eymerich's method followed the concepts established in the gloss on *Quod super nonnullis* and *Super illius specula* that were in turn used by earlier inquisitors. It depended less on determining whether the art in question was naturally possible than on investigating the behavior of the person by whom it was practiced. For Eymerich, it was the inclusion of ritual elements during the performance of an art that signaled that its practitioner had crossed the line between legitimate and illegitimate behavior. Aquinas would certainly have agreed that an activity that included ritual elements designed to invoke demons was unacceptable to Christians. Nevertheless, Eymerich's focus on the diviner's behavior rather than the validity of the art itself led him to overlook a third kind divination identified by Aquinas. As we have seen, Aquinas had noted that there were also activities that were ostensibly natural but that nevertheless operated solely with the connivance of demons. He had condemned predictive arts such as geomancy, haruspicy, and chiromancy on the grounds that there was no natural basis for the predictions that they produced. Since such arts did not include any ritual elements, they would not necessarily have been included in Eymerich's account of "heretical" divination. It is also notable that in this section of the *Directorium* at least, Eymerich had relatively little to say about the legitimacy of predictive astrology, an issue that Aquinas had explored in detail.[53]

The disparities between Eymerich's and Aquinas's ideas may be explained by the fact that the former engaged with only a relatively small portion of the Thomist corpus when writing this section of his handbook. When Eymerich invoked Aquinas's authority, he cited texts such as his commentary on the *Sentences* rather than the *Summa theologica* or *Summa contra gentiles*. In fact, in the two questions of the *Directorium inquisitorum* concerned with magic, Eymerich referred to the *Summa theologica* only once. Moreover, when he did cite it—during his discussion of invocation—he did so only indirectly, when quoting passages of legal opinion that had used this text. Consequently, Eymerich's handbook did not directly address the arguments that Aquinas had formulated in his discussion of superstition in questions 92–96 of the *secunda secundae* of the *Summa theologica*. In these sections, Aquinas had developed fine-grained distinctions to assess the intellectual validity of particular genres of magical practice. These included the argument that Christians could perform wonders not only by making explicit pacts with demons—for example, through the performance of ritual—but also unwittingly through the conclusion of "tacit" pacts. Although Eymerich did not dismiss the possibility of tacit pacts, he did not discuss them in any detail.[54]

Eymerich's interest in identifying individuals who explicitly rendered dulia and latria to demons also informed a later section of the *Directorium inquisitorum* in which he offered advice to inquisitors about how to proceed against alleged heretics. It included a section entitled "The Signs by Which a Heretical Necromancer Is Distinguished," in which he argued that heretics shared certain distinctive marks that could be identified by the inquisitor. Recalling the principles of physiognomy, these included aspects of the magician's physical appearance. Other signs related to the magician's activities. Eymerich noted that a further useful indicator of a suspect's guilt was whether they also happened to practice astrology or alchemy. Although he did not consider these activities to be necessarily heretical, he maintained that those who engaged in them often resorted to necromancy, "for when they cannot attain their intended ends, they ask for the assistance of demons, they invoke and they implore them, and by imploring they conjure and sacrifice either tacitly or expressly."[55] While Eymerich did not here offer a developed account of these operative arts, it was consistent with his earlier account of divination. He appeared to condemn the activities of those who would practice "mixed" forms of alchemy or astrology—that is, those who combined seemingly legitimate practices with forms of invocation. In this instance, however, he did appear to concede, in line with Aquinas, that invocation need not always by explicit.[56]

In his detailed discussions of magic and divination, Eymerich presented a broadly consistent approach to these arts. Elsewhere in his handbook, he introduced an element of ambiguity. In part 2, questions 23–29, he reproduced a series of lists of banned books. Some were taken from papal decrees and collections of canon law. They included the bull *Conservationi puritatis*, which purported to have been issued by Gregory XI and which condemned twenty of Raymond Lull's works and extracted two hundred propositions from his works that were considered heretical. In another entry titled "Books Condemned by Inquisitors of Aragon," he included a number of works by Arnald of Villanova, though none specifically concerned with magic. In the final entry, "Books Condemned in Italy and France," he included a reference to a prohibition effective in Paris (*in partibus Gallicanis, videlicet Parisiis*) drawn up by the bishop of Paris and the Dominican inquisitor, which had banned all works treating divination and sortilege, including works of necromancy, hydromancy, and chiromancy. In each of these works, the entry continued, the authors described arts that involved either using incantations or making tacit and manifest pacts with demons. The date of this original decree is unclear, but the content certainly indicates the influence of learned discussion of the boundaries of nature that had occurred during the thirteenth century. The specific bans on arts such as chiromancy nevertheless also contradicted Eymerich's own assessment of their legitimacy and called into question the basis on which he had arrived at this judgment. As we shall see in chapter 7, Eymerich's decision to include both Gregory XI's bull against Lull and the list of arts banned in Paris would create significant difficulties for clerics attempting to interpret and use his work in the sixteenth century.[57]

The Pursuit of Magic at the Close of the Fourteenth Century

The overall numbers of sorcery trials began to climb across Europe in the last quarter of the fourteenth century. In absolute terms the increase was slight, rising to between one and two trials per year across Europe—that is, a figure comparable to that seen at the beginning of the century. Kieckhefer argued that these increased numbers were probably influenced by the introduction of the inquisitorial method into municipal courts. His point was supported by the fact that most of the trials were carried out in secular rather than ecclesiastical tribunals. It is, however, telling that civic authorities were, in accordance with papal desires, directing their new techniques against forms of superstition. The prosecutions were also spreading to new areas. For example, there were no known prosecutions for sorcery in Switzerland before 1383, but in the closing decades of the fourteenth century there were four minor cases. They were followed by a significant series of trials held in the Simme Valley circa 1395–1405. Conducted by the secular judge Peter of Greyerz, they led to the execution of an unknown number of people for sorcery. These mass trials for sorcery were unprecedented, and there were no further examples elsewhere in Europe at this time.[58]

These patterns of prosecution were also broadly consistent with events in Italy. In Florence, the first extant civic trials for sorcery begin to appear in 1375. In that year, a woman named Catherina was accused of having enchanted a Florentine merchant by placing a wax image in his bed and making certain incantations. Some twenty years later, Nicolosa Vanzi was examined for producing a charm to make her neighbor fall in love with her. In a further case, Jacopo di Francesco was charged with various crimes including seeking to perform love magic by writing a psalm on a slip of paper and leaving it where it could be picked up by the object of his desires. He also performed magic on behalf of others—for instance, a woman requested that he produce a locket inscribed with characters in the hope that it would allow her to fall pregnant. It is unclear why the Florentine authorities began to investigate these cases, but Bruckner suggested that it may be connected to the establishment of a new political regime less tolerant of superstition and heresy. This may well be true. The willingness of political elites to act against sorcery nevertheless needed to be informed by a conception of what constituted illicit practices. It is striking that the forms of magic investigated were consistent with those that the papacy had been seeking to address.[59]

Italy appears to be unique at this time, because it seems to have been the only region in which inquisitors were investigating sorcery. As we have seen, the first recorded inquisitorial trial for magic occurred in 1360 and a second in 1370; a further five took place in Italy before 1400. The first occurred in 1383, when a Tuscan inquisitor—likely to have been Fra Piero di Ser Lippo—wrote to the government of Siena to inform them that he had discovered a group of sorcerers in the village of Rugomagno. A certain Agnolo di Corso had been circulating a book of magic that others read and copied. Using this work, he continued, the villagers invoked demons in order to work maleficiant magic. A year later, Fra Piero heard the case of one Niccolò Consigli in the city of Florence. The records show that the accused had been summoned before the court some years previously, answering charges that he had attempted to perform an exorcism on a young girl. He was summoned to appear once more before the court in 1376, but, having failed to do so, he was excommunicated. Consigli was finally captured in 1384. While imprisoned, he refused confession and pledged himself to the devil. The inquisitors found that he had regularly practiced forms of sorcery and divination that were variously intended to inflict maleficia, perform exorcisms, find lost property, foretell the future, or provoke either love or hatred in selected individuals. More importantly, however, they determined that in order to work these effects he had made incantations to and conjured demons, precisely the activities that the papacy had suggested were to be treated in the manner of heresy. Consigli was sentenced to death for his crimes and executed in Florence.[60]

These two cases suggest that by the later fourteenth century, inquisitors active in parts of Italy had begun to act on the papacy's encouragement to investigate cases of heretical magic. Since the inquisitorial records for Tuscany do not survive—these documents happened to be preserved in the records of the podestà—it is impossible to assess how common it was for inquisitors to prosecute such cases. There is, however, evidence to suggest that contemporaries considered the inquisitors' actions to be unusual and indeed unwelcome. In his letter to the Sienese government, the Tuscan inquisitor lamented that certain citizens of Rugomagno had demanded that Angelo di Corso should not be punished. Furthermore, an anonymous chronicler who described Consigli's execution suggested that the local bishop, clergy, and doctors of canon law were opposed to the execution. Although inquisitors had evidently been investigating at least some cases of sorcery in Tuscany for several years—Consigli was first investigated before 1376—the penalty meted out in this case appears to have been considered unusual and a cause for concern. These two cases may be the product of the enthusiasms of one inquisitor, but equally those enthusiasms may have been inspired by the papacy's efforts to encourage inquisitors to investigate cases of sorcery conducted by demonic means and punish them in the manner appropriate for heretics.[61]

The magical practices that inquisitors encountered in the later fourteenth century did not, however, always conform to those described by the papacy or recorded in inquisitorial handbooks. In 1384, two women, Sibillia and Pierina, were separately placed on trial in Milan for attending nocturnal meetings. As Kieckhefer has convincingly argued, these images of strange nocturnal rites constituted a distinctive mythology of witchcraft sustained in parts of northern Italy. Taking place every Thursday, each meeting opened with the women bowing their heads in greeting to a woman named Oriente who presided over the society. During the meetings, Oriente would reveal to the assembled company privileged information about matters future or unknown. The women also believed that Oriente could revivify dead animals. As in the case of Benvenuta, discussed above, neither woman believed that she had done anything contrary to the faith. Both women abjured their beliefs and were released, but six years later each was rearrested and admitted reverting to her old habits. They were convicted as relapsed heretics and sentenced to death.[62]

The beliefs that the inquisitors encountered in Milan did not readily conform to their conception of magic, but it seems that they nonetheless began to translate them into their own terms of reference. From the first trial, inquisitors referred to these meeting as the "Games of Diana," seemingly invoking the games described in the canon *Episcopi,* and this document may provide a key to interpreting the inquisitors' reaction to this case. The canon held that the "Riders of Diana" had been seduced by demons and fooled into believing that they took part physically in the games. The sense that the women were deluded was also suggested by their belief that Oriente could bring creatures back to life, a feat that the canon intimated was impossible. The trial documents indicate that the inquisitors were nevertheless seeking evidence of the women's complicity in the devil's works. In 1384, the charges against Pierina included, "Also you said this Mistress [Oriente] taught you of the society the virtues of herbs and also by means of signs that you gave to her she revealed to you everything that you asked of her regarding sickness, theft and maleficia."[63] The reference to offering signs appeared to indicate that her activities involved offering reverence to demons. This accusation was more clearly articulated in the summary of her 1390 trial, which recorded that "when she wanted to go to this game she called the spirit Luciffelus who always came to her." This spirit, who always appeared in the shape of a man, also "talked with her and instructed her about everything that she asked of him."[64]

It is perhaps a reflection of the growing influence of the conception of magic advanced by the papacy that while inquisitors may have doubted the reality of the games described by Sibillia and Pierina or at least the truth of some of the events that transpired there, by the time they executed them they were convinced that they were soliciting the assistance of demons for reasons and by means familiar to them. These views do not appear to have been universally shared, however. As we have seen, Gregory XI had suggested that there was resistance to papal inquisitors' authority to prosecute magic in France in the 1370s. This appears to have continued, for, in 1398, theologians at the University of Paris returned once more to the question of whether acts of magic or sorcery that involved idolatry and apostasy were tantamount to heresy. At stake once more was the question of an inquisitor's jurisdiction over magic. Although the theologians agreed with the principles established by the papacy, the fact that it was necessary for them once more to reaffirm this point underlines how contentious this jurisdictional claim remained.[65]

Conclusion

By the end of the fourteenth century the papacy had reasserted its right not only to appoint inquisitors but also to authorize them to prosecute magic. The papal bull *Super illius specula* established that inquisitors could prosecute practitioners of magical arts that involved directly invoking demons or corrupting Christian religious practice in the manner of heretics. Such a position ignored the detailed and sustained analysis of arts that contained no ritual element carried out by leading church intellectuals such as Albertus Magnus and above all Thomas Aquinas. The papacy's new definition neither significantly increased the prosecution of magic nor gained universal assent. By the 1370s, Nicholas Eymerich still felt the need to justify the inquisition of magic. He bolstered the central argument from *Super illius specula* with selected arguments from Aquinas's corpus to construct a new hybrid definition of magic. Aquinas's detailed account of superstition contained in the *Summa theologica*, including his critique of putatively natural forms of magic, was at most marginal to Eymerich's account. Furthermore, the prosecution of magic remained rare in all courts. This situation was soon to change. As we shall see in the following chapter, over the course of the fifteenth century, Dominican intellectuals began to recover Aquinas's detailed account of superstition and to prosecute an unprecedented number of cases of magic.

II

Mendicant Reform and the Inquisition of Magic

4

The Crisis of Papal Authority and Observant Reform, 1378–1500

Introduction

The pursuit of magic intensified dramatically during the fifteenth century. This change was connected to a crisis of papal authority that began with the Great Schism and resulted in the conciliar crisis of the fifteenth century. The conciliar crisis exposed long-standing disputes within the Church about the respective rights and authority of the papacy and the episcopate, while raising questions about the most appropriate means to define and organize reform. Although the papacy eventually fended off the conciliar challenge, its authority was severely damaged, and the ideas formulated by the conciliarists continued to present a potent threat. Responding to conciliarism, the papacy enacted measures to rebuild its shattered authority. Most notably for our purposes, the papacy encouraged the mendicant Observant reform movement. With papal support, the friars embarked on a renewed program of pastoral reform that stimulated a renewed campaign against superstition. This new program coincided with an intellectual revival within the Dominican order that prompted two profound changes in thinking about magic. The first, the development of the elaborated stereotype of the witch, is well known; the second, a revival of Thomist thought on superstition, is not.

The Great Schism and the Rise of Conciliarism

In the mid-1370s, the medieval papacy remained at the height of its influence. Its power was undimmed by its relocation to the city of Avignon; indeed, in some respects papal authority had been enhanced during this period. Now permanently resident in Avignon, the papal bureaucracy enjoyed greater stability, which enabled greater centralization. Notably, this development enhanced the papacy's power centrally to appoint bishops and abbots. In principle, this development allowed them to promote reform, but it also had the effect of encouraging a turnover in benefices to allow the papacy to collect taxes. The growth of the papal bureaucracy also enhanced the power, authority, and wealth of the cardinalate.[1] The Avignon system was disrupted in 1377 when Pope Gregory XI (1370–78) moved his court from Avignon to Rome. After his death a year later, the cardinals elected Urban IV (1378–89) as his successor. Unexpectedly, the new pope enacted reforms designed to curb the power and privileges that the cardinals had enjoyed under the Avignon system. Angered, a number of the cardinals elected a second pope, Clement VII (1378–94). Supported by the French king, Charles VI (1380–1422), Clement established a rival papacy based in Avignon and sought to end the crisis with military force. These events marked the beginning of the Great Schism.[2]

The initial crisis came to a head in 1408 when cardinals drawn from both the Roman and Avignon obediences withdrew their support from their respective popes and sought to organize a council in Italy to elect a replacement. Their actions were given legitimacy by the Parisian theologian Jean Gerson (1363–1429), who justified the subtraction of obedience from a pontiff by building on the arguments first developed by secular theologians during their dispute with the mendicants in the thirteenth century. The secular theologians had maintained that since Christ had directly established the episcopate, the papacy did not enjoy absolute authority over his fellow bishops.[3] The council opened in Pisa on March 25, 1409, but rather than resolving the schism, the council exacerbated the crisis by creating a third rival papal line. In an attempt to resolve this new crisis, Emperor Sigismund pressured the Pisan pope, John XXIII, to hold a further council, which opened in 1414 in the imperial city of Constance.[4] The meeting almost collapsed in March 1415 when John XXIII fled to Schaffhausen, but it was saved by the intervention of Gerson, who was attending as part of the French delegation. Addressing the assembled fathers, he asserted that since councils were directly instituted by Christ, they possessed authority over all Christians, including the pope. In the wake of his intervention, the fathers began to adopt a more assertive position. During the council's fifth session (April 6, 1415), they issued the decree *Haec sancta synodus*. It declared that universal councils derived their authority directly from Christ and that they could exercise authority over popes "in matters of faith, unity and reform."[5]

By 1417, the council had deposed each of the rival popes. Before electing a replacement, the council fathers passed the decree *Frequens* (1417). It stipulated that, henceforth, popes should regularly convene ecumenical councils, which they argued would enable the papacy to restore and renew a neglected Church.[6] In November 1517, the fathers elected Oddo Colonna (1369–1431) as Martin V.[7] Having established his court in his native Rome, the new pope signaled his commitment to Church reform by passing statutes reducing the size of the curia, restricting papal rights of taxation, and curtailing the papacy's rights to bestow benefices. Fulfilling the requirements of the decree *Frequens,* he convoked a further council to be held in Basel in 1431. Since Martin V died before it could open, the Council of Basel was overseen by his successor, Eugenius IV.[8]

The council was initially poorly attended. Eugenius, who feared the council's power, used this fact as a pretext to demand its dissolution. The delegation from the University of Paris, led by Gerson, rebuffed this demand and reaffirmed the decree *Haec sancta.* In February 1432, the council fathers reinforced its provisions by passing a further decree, *De stabilimento concilii,* which stated that a council could not be transferred, prorogued, or dissolved without its consent.[9] These declarations marked a new stage in episcopal efforts to redefine their authority with respect to the papacy. Having marginalized Eugenius, the council fathers began to develop their own program of reform. The council's early reform decrees centered on the office of the bishop. The fathers enacted measures designed to address venerable concerns such as clerical celibacy and simony but also decreed that prelates should be elected locally and that bishops should hold regular diocesan councils. These reforms reflected a new emphasis on the sacramental and pastoral dimensions of the bishop's role while also stressing their autonomy. They became a model for bishops to follow during the later fifteenth century and subsequently informed the reforming agendas of the fathers of Lateran V and Trent.[10] More controversially, and indicating their desire to exert an influence over the papacy, the council fathers also proposed reforms to the "head" of the Church, specifying, for example, the number and quality of the cardinals.[11]

In 1437, Eugenius was presented with an opportunity to regain the initiative over the council. The Greek emperor and patriarch had indicated to the council that in order to gain the support of the Latin West in their conflict with the Ottomans, they would be willing to discuss their theological differences. The Greeks were unwilling to travel beyond the Alps, however, so Eugenius proposed that the council should be moved to Italy. Translating the council also offered Eugenius the chance to reassert his authority, a fact recognized by the majority of the council fathers, who chose to remain in Basel. The dispute between pope and council subsequently deepened. The fathers declared that the council was unequivocally superior to the pope and that he could not be dissolved or transferred without its consent. In 1439, they went so far as to depose Eugenius and appoint a new pope, Felix V, thus starting a new schism.[12] During this confused period, a number of secular rulers adopted a position of neutrality while considering the decisions already taken at Basel. Notably, in France prelates voted to accept the decrees issued at Basel, including the general principle of the supremacy of councils. On July 7, 1438, King Charles VII (1403–61) translated these decisions into secular law in the Pragmatic Sanction of Bourges. Its passage marked the culmination of French monarchs' efforts to limit the influence of the papacy within their realm that dated back to the reign of Philip IV.[13]

Eugenius died in 1447, leaving the crisis unresolved. To bring it to a close, his successor, Nicholas V (1447–55), was forced to make a series of compromises. He agreed to a number of concordats with secular rulers to secure their obedience, including accepting the terms of the Pragmatic Sanction. The Council of Basel finally ended in 1449. Nicholas signaled his willingness to reach an accommodation with the conciliarists by recognizing the papacy's obligation to call another meeting within eighteen months. At the same time, Nicholas worked to restore papal power. Within Italy, he sought to secure the territorial integrity of the Papal States and to place not only the papacy but also the city of Rome at the center of the Catholic faith by investing in monumental building works and patronizing art and literature.[14] These efforts were reinforced by efforts to reestablish the theoretical basis of papal authority. In his work *Summa de ecclesia*, the Dominican theologian Juan de Torquemada (1388–1468) rejected *Haec Sancta*'s legal and dogmatic validity. He instead articulated an ecclesiology that posited that the pope, as Peter's successor, possessed all authority within the Church, which he in turn delegated to bishops. He also used this principle to defend the papacy's right to grant privileges to the mendicants to undertake the cure of souls. Addressing the power of councils, he argued that they had been instigated by Peter rather than Christ and denied that they possessed any authority other than that given to them by the pope. It was therefore impossible for council fathers to confer on themselves any power over a pope. In 1460, Pope Pius II consolidated these arguments by passing the bull *Exercrabilis*, which forbade an appeal to a council over the head of a pope.[15]

Observant Reform and the Pursuit of Superstition

The Observant reform movement began in the later fourteenth century when a group of Franciscans sought to create reformed houses that would accommodate friars who wished to observe a stricter interpretation of the order's original rule. The adherents of the new movement became known as the "Observants," a title that distinguished them from unreformed "Conventual" friars. Members of the Augustinian order initiated a similar reform in 1386, followed in 1388 by a number of Dominicans and members of other monastic orders, such as the Benedictines.[16] During the fifteenth century, Observant reform in Italy expanded rapidly and entered a new pastoral phase. These latter developments were related to the recent upheavals that had rocked the Church and gravely undermined papal authority.

During the period of the schism and subsequent conciliar crisis, the laity had begun to play a more significant role in religious life in Italy. Individual Christians participated more actively in their faith through joining confraternities or directly engaging with scripture and the works of the church fathers. For their part, civic leaders began to enjoy greater influences on local churches.[17] Some, such as Cosimo de' Medici (1389–1464), encouraged Observant reform in the peninsula by offering practical and financial support for the development of new religious houses.[18] Civic authorities also invited preachers such as the Dominican Giovanni Dominici (1356–1420) and the Franciscan Bernardino of Siena (1380–1444) to address urban populations. After the closure of the Council of Constance, Martin V and his immediate successor, Eugenius IV, sought to restore papal authority by giving the friars, who owed their allegiance directly to the papacy, enhanced responsibility over the reform. For example, in 1446, the leader of the Dominican community at St. Marco, Antonino Pierozzi (1389–1459), was appointed to the role of archbishop of Florence by Eugenius IV. Echoing the aspirations for episcopal reform expressed at the Council of Basel, Pierozzi resided within his diocese and personally visited its parishes.[19]

During this period, the mendicants began to expand their original ministry, by preaching and hearing confessions of urban populations. By these means they aimed to eliminate deviant or superstitious behavior while simultaneously encouraging what Roberto Rusconi has termed positive "behavioral models." Making full use of the high levels of literacy in the towns of northern and central Italy, the friars harnessed the power of the printing press to produce vernacular printed confessional works intended for a lay audience, which enjoyed a wide circulation in the last in two decades of the fifteenth century.[20] Some of the friars' efforts reflected the particular circumstances of fifteenth-century Italy, for they warned of the dangers of moral deviance and materialism caused by an increasingly wealthy society. Bernardino of Siena went so far as to organize bonfires of vanities, during which he exhorted citizens publicly to dispose of the most egregious symbols of their wealth. The friars also continued to address areas of long-standing concern. To take one important example, they continued to preach against the Jews, and their sermons frequently condemned the Jews' involvement in usury. In 1462, Franciscan friars in Perugia proposed novel means to ameliorate the perceived danger of Jewish moneylending by establishing the first *montes pietatis* loan system.[21]

From the 1420s, efforts to purge Christian society of superstition became an increasingly important aspect of Observant reform. In a famous sermon delivered in Siena in 1427, Bernardino of Siena urged its citizens to act against practitioners of magic. To make his case, he recounted his involvement in the prosecution and execution of an old woman in Rome three years earlier. Without being tortured, the suspect, Finicella, had confessed to killing thirty children by sucking their blood. She also admitted to killing her own son and producing a powder from his body for others to eat. Having recounted this grisly story, Bernardino exhorted the people of Siena to "burn a little incense to God"—that is, to seek out and execute other witches like Finicella.[22] On this occasion Bernardino's audience did not heed his call to action. Later that same year, however, he also preached in the city of Todi. The following year, a woman named Matteuccia was tried in Todi's civic court for perpetrating crimes similar to those described in Bernardino's sermon and was subsequently found guilty and executed.[23]

It is important not to overemphasize either the apparent novelties in Bernardino's preaching or their immediate impact on the development of the pursuit of superstition in Italy. When Bernardino preached the sermon of 1427 in which he related the story of Finicella, he referred to her as a *strega*, a word that is often translated as "witch," and there are apparent similarities between Finicella's actions and those of stereotypical witches. He not was not, however, seeking to condemn practices consistent with those denoted by the historiographical category "witchcraft." As Richard Kieckhefer has convincingly argued, we should not interpret Bernardino's beliefs about the *strega* as a local manifestation of a mythology of diabolic witchcraft that was developing simultaneously across Europe. He instead suggested that Bernardino's idea of the *strega*—like the complex of beliefs surrounding Oriente that we encountered at the end of the last chapter—derived from a distinct, localized mythology of witchcraft. Rooted in ancient Roman culture, this mythology was specific to parts of Italy, and it held that certain women entered houses by night and killed children.[24]

Furthermore, Bernardino's preaching about the danger posed by *streghe* formed only a part of a broader set of concerns about superstitious magic. Many of his ideas were similar to those of his fellow friars and the papacy. During a sermon delivered in Assisi, for example, he warned his audience, "If you know of some diviner, or interpreter of dreams or augur, and they truthfully say, as it is damned to intend to say something contrary to the truth, that they know something through divination and moreover they know it through diabolic incantations, and if the thing of which they speak should come to pass, they do it to draw you into errors of the Christian faith and to make you worship idols."[25] Bernardino's words echoed Augustine's warning that the diviner's arts were false and functioned by demonic means and that those Christians who listened to their predictions were themselves drawn into superstition. He adopted a similar approach while preaching against the practice of making talismans, in this context probably breve or slips of paper inscribed with some form of incantation or prayer rather than more elaborate astrological talismans. He stated that "they sin all those who write [talismans], those who carry them, those who give them, those who sell them, those who buy them, those who teach carrying them, those who believe them to be good."[26]

What impact did the mendicants' preaching have on the prosecution of magic? At the beginning of the fifteenth century, it appears that, if anything, the inquisition of magic became less common in Italy than it had been in previous decades. Lists of trials supplied by Kieckhefer and Andrea Del Col show that between 1400 and 1420 there were only four trials for magic in Italy, which resulted in the deaths of six people. None were heard in an inquisitorial court.[27] The decline of prosecutions for magic may be more apparent than real, a misleading impression created by the absence of evidence, but the existing data suggests that during the opening years of the fifteenth century the inquisition of magic was not a major focus of mendicant activity. Possibly encouraged by mendicant preaching, prosecutions for magic rose slightly. During the 1420s there was a total of seven trials, two conducted by inquisitors, which resulted in six executions. The relatively stable figures suggest that although Bernardino's preaching may have inspired some prosecutions, they did not trigger significant outbreaks of witch-hunting. This may be because many in the Church resisted these ideas. As Bernadette Paton has shown, contemporaries such as the Augustinian friar Andrea Biglia were critical of both Bernardino's beliefs and his zeal for prosecutions.[28]

Although Biglia's objections show that there was some resistance to the pursuit of magic among lower-ranking clerics, the papacy continued to encourage its prosecution. In 1437, at the height of the conciliar crisis, Eugenius IV wrote a letter addressed to all inquisitors of heretical depravity informing them that it had come to his attention that many people had made pacts with the devil in order to work effects and create maleficia. Michael D. Bailey has suggested that in this letter Eugenius was implying the existence of "an organized conspiratorial cult headed by Satan himself." In effect, his letter provided an early intimation of the stereotype of the witch.[29] While this is possible, the prosecution of two women in

Milan in 1380 (discussed in the previous chapter) showed that it was possible for ecclesiastical authorities to believe in the existence of groups of people who gathered to offer reverence to demons in order to work effects, without necessarily subscribing to the elaborated stereotype of witchcraft. Moreover, in the remainder of the letter, Eugenius's concerns were similar to those of his predecessors. He claimed that certain individuals were sacrificing to demons with the intention of inflicting or lifting maleficia, curing illnesses, or creating storms. He also expressed the fear that to carry out their malign acts, these magicians were abusing the sacraments of baptism and the Eucharist. Others, he complained, made and used "images of wax or other things with invocations or, having created things of this sort, they baptize them or order them to be baptized." Rather than wanting inquisitors to prosecute a heretical sect of devil-worshippers, Eugenius appears to have been encouraging them to redouble their efforts to extirpate forms of superstition known to the papacy since the fourteenth century.[30]

Despite the efforts of the papacy and the mendicants to encourage the pursuit of superstition, rates of prosecution remained steady in the 1430s. There was a total of six trials, of which only one took place in an inquisitorial court. These various prosecutions resulted in eighteen executions. Although the number of prosecutions did not increase, there is evidence that the concept of superstition promoted by both the papacy and the friars was continuing to inform the actions of secular tribunals. In the city of Perugia, for example, magistrates prosecuted a series of trials for magic in the mid-fifteenth century. One set of documents concerned the case of a certain Giacomo di Nicolò Cervi da Pisa, who in 1433 was found to have fraudulently claimed to be able to work magic. He was accused of convincing a prefect of the city who was suffering from an unspecified infirmity that he knew how to relieve his symptoms by producing a talisman inscribed with what he claimed were incantations and teaching the prefect how to recite certain incantations. The document added, however, that "in truth, he did not perform enchantments and nor did he know how."[31] In a further example of Giacomo's wrongdoing, the documents record that in order to prove he could work magic, he encouraged a group of men to visit a certain house. While there, Giacomo placed himself in a circle and recited incantations, but as the document tersely adds, "he pretended and craftily showed that he knew how to recite incantations in the abovementioned manner when in truth he knew nothing." Despite the magistrate's skepticism, the case reveals that Giacomo and those he sought to dupe shared a common understanding of magic, what it could be used to achieve, and in principle how it ought to function. Each recorded act involved making an explicit appeal through the performance of ritual or incantation to a demon.[32]

The Perugian magistrates did not always dismiss charges of magic. Trial documents also record the case of Caterina di Giorgio da Modrus, who was charged with a variety of offenses including working love magic on behalf of an unnamed man. To produce her effect, she "approached a certain priest, whose name it is better not to mention, and by a subtle trick sought to have, and obtained, from him a certain consecrated host, and on this host Caterina wrote certain characters and the names of certain demons." Caterina took the host and other items, including coals burned on Christmas day, and tied them up in a piece of linen. She placed the resulting parcel by the threshold of the house in which lived the woman whom the unnamed man desired. Once again, the magistrates were not simply concerned by the fact that Caterina had performed a maleficiant act but that she had done so by working a form of magic that involved explicitly appealing to a demon. The aggravating factor in this act of love magic was the abuse of the consecrated host obtained from a priest by deception. Each of these actions had been specifically condemned in documents such as *Super illius specula*.[33]

Evidence from Perugia also suggests that here, too, it was possible for concerns about superstitious magic recognizable to ecclesiastical authorities to mingle with local mythologies of witchcraft. In one of his sermons, the Franciscan preacher Giacomo della Marca (1391–1476) mentioned the case of a woman known as "Santuccia," who was burned at the stake in 1445 for killing no less than fifty boys by sucking their blood. In addition to these crimes, he noted that "she had performed many maleficiant acts with the body of Christ consecrated by her friend a priest."[34] In a further case, Fillipa da Città della Pieve was charged in 1455 with a number of offenses, including working powerful forms of love magic. The magistrates found that Fillipa had learned many of her arts from another woman, Clarutia, some twenty years earlier. On one occasion, under her guidance, Fillipa had formed a pact with a demon by stripping naked, applying an unguent, and promising herself to the devil. In return, a demon transported the women to a neighboring house, where they sucked the blood of a one-year-old child, causing his death. The women subsequently removed the child's body from the grave and used it to produce further magical potions. As Kieckhefer has shown, contemporaries appear to have regarded these ideas not as the introduction of a novel concept of a witch but as a familiar and comprehensible part of their belief system.[35]

These cases have demonstrated that the Perugian magistrates' concept of superstition was consistent with the ideas set out by the papacy in *Super illius specula* and reiterated by Eugenius IV. They also echoed the ideas set out by the friars—for example, by Bernardino of Siena in his preaching and in inquisitorial manuals such as Eymerich's *Directorium*. Each of the Perugian trials turned on the question of whether the accused had committed—or had at least claimed to be able to commit—specific acts explicitly recognized to involve making an explicit appeal to a demon. While there were occasional references to *streghe*, there is no evidence for the early spread of the elaborated stereotype of the witch. There was another point of similarity. None of the Perugian trials examined whether the accused had performed an art that he or she claimed functioned by natural means, such as predicting the future through physiognomy or seeking to channel astral influence by constructing astrological talismans. Absence of evidence is not—as the cliché goes—evidence of absence. We cannot definitively conclude from these trials that neither civic nor religious authorities were interested in investigating forms of ostensibly natural magic. This pattern might simply reflect the nature of the cases brought to trial in Perugia. It is, however, also likely that the papacy's emphasis on identifying evidence of ritual practice had created a situation in which both ecclesiastical and civic authorities were relatively uninterested in ostensibly natural operative arts. As we shall see, this situation began to change during the remainder of the fifteenth century.

Johannes Nider, Witchcraft, and Superstition

From the 1430s, a number of texts began to describe a heretical sect that gathered to worship Satan, denigrate the Christian faith, sacrifice children, and conspire to commit maleficiant acts. The origins of this new complex of beliefs remain obscure. One of the most influential accounts of the new stereotype of the witch was contained in the *Formicarius*, a work written by the Observant Dominican Johannes Nider (1380–1438). Neither an inquisitorial manual nor a learned treatise on magic, as Michael D. Bailey has shown, it was designed for the use of preachers, offering them a series of edifying stories that they could use to exhort Christians to engage in moral reform. It was divided into five parts, of which the first four described the deeds of good men and women, revelations, false visions, and the saints and holy men. Only the fifth part was specifically concerned with magic and witchcraft. It related a series of lurid tales, which contained several of the ideas that would subsequently inform the stereotype of the witch. Nider claimed that he had learned of the new sect of witches from several sources, including a former necromancer, a Dominican inquisitor of Autun, and the accounts of Peter of Bern, who had prosecuted cases of magic in Switzerland during the early 1400s. Nider suggested that these earlier trials had been conducted to disrupt an organized sect of witches, but it is likely that he was projecting the novel concept of witchcraft that he was describing in the *Formicarius* back onto these earlier investigations.[36]

The long-standing historiographical aim of pinpointing the causes of the witch hunts has ensured both that the *Formicarius* remains Nider's most studied book and that his contribution to the stereotype of the witch is his most well-documented legacy. This tendency has obscured his role in a further, equally profound transformation in the conceptualization of magic in the fifteenth century. Other historians have previously noted that in his *Praeceptorium divinis legis* Nider drew on Aquinas's work, especially the *Summa theologica*, to consider magic, but they have not explored in detail the implications of this development.[37] Nider's discussion of superstition was directly shaped by his reading of the relevant sections of the *secunda secundae* of the *Summa theologica*, which, as we have seen, had not hitherto been used to inform the prosecution of magic. The specific reason why Nider chose to engage with these materials is unclear, but his decision was likely to have been informed by a revival of interest in the writings of Aquinas that was becoming evident in both the universities and the religious orders during this period. Notably, by the mid-fifteenth century, the *Summa theologica* had replaced Peter Lombard's *Libri quattuor sententiarum* as the basic text of theological instruction in a number of German convents, and it gradually began to be taught in other convents and then in the universities.[38]

Reading the *Summa theologica* led Nider to articulate a means to determine the boundaries of illicit superstition that differed significantly from the one proposed in *Super illius specula* and *Directorium inquisitorum*. These influential documents had emphasized the identification of ritual activity, thereby diminishing the significance of arts that involved no explicit ritual element. While Aquinas had condemned any art that required its operator to make an explicit pact with a demon, he had also questioned the legitimacy of arts that functioned by ostensibly natural means. He had concluded that any art that did not produce its effects from demonstrably natural causes was likely to depend on the assistance of demons. Nider appears to have accepted these principles. In one telling passage, he noted that "someone can strive to investigate hidden or future things, either by means of an occult or manifest pact that they had entered into with demons."[39] While this reference to an occult or tacit pact may appear insignificant, it reflected an important conceptual reorientation in his work that was informed by Thomist principles. By reemphasizing the idea of a tacit pact, Nider encouraged demonologists, inquisitors, and magistrates to question whether a particular operative art was naturally possible rather than simply whether it involved the explicit invocation of demons.

Recovering these Thomist principles encouraged Nider to consider in greater detail the natural order and the extent of human and demonic power therein. This new approach was evident when he addressed the questions: "Whether it is true that witches can make real animals by their art?" and "Whether by the work of demons transformations of men can be made into beasts?" He began his answer to the latter question by considering the opinions of Augustine before remarking that "Thomas, however, in the [*Summa theologica*] *pars prima* q.114 a.4 says that all transformations of bodily things, which can be made by means of certain natural virtues, to which they belong seeds which are found in the elements of this world, can be made by the operations of demons, having employed seeds of this sort." In this passage of the *Summa*, Aquinas had drawn on Augustine's authority to argue that demons could transform objects into serpents or frogs. Nider concurred with his view, thereby acknowledging that it might be possible for demons to effect wondrous transformations. Nider thus endorsed the broad Thomist principle that demons could create wonders, but they could only do so naturally—that is, by means of art. He therefore also accepted Aquinas's further conclusion that demons could not perform any naturally impossible acts, such as restoring the dead to life.[40]

Aquinas's influence on Nider's conception of superstition was also evident in his discussion of specific operative arts. In question 38, for example, he discussed predictive astrology. To do this he cited the *Summa theologica* (2.2 q.95 a.5), offering a tight summary of Aquinas's ideas. Nider agreed with Aquinas's assertion that it was possible to make some forms of knowledge from the stars. Echoing Aquinas, he noted that "they can be known certainly, those things which occur necessarily through the movement of the heavens."[41] Nider was referring here to events such as eclipses. It was, he continued, nevertheless impossible to predict two further classes of event. The first was chance events or those that occur accidently, for, he noted, following Aquinas, accidental events "do not have a cause, especially a natural cause."[42] Second, he argued that one could not use astrology to predict any acts that depended on the human will, for as Aquinas had explained in the *Summa theologica*, corporeal beings such as the stars could not affect the incorporeal will. Consequently, if an astrologer claimed to have made a certain prediction about human actions, he or she must have been assisted by a demon. Nider nevertheless reaffirmed the position advocated by Aquinas that it was possible for astrologers to make uncertain predictions about future human actions, on the ground that the stars could incline human behavior. Although unoriginal, Nider's line of reasoning was significant because he was using the criteria that Aquinas had developed in the *Summa theologica* to determine the boundaries of superstition. Rather than simply advocating the examination of the actions of the person practicing a given art, Nider now encouraged the consideration of whether that art was naturally possible.[43]

The New Mythology of Witchcraft in Italy

The new concept of witchcraft developed in works such as the *Formicarius* had a significant impact on the inquisition of magic in western Europe. The first trials inspired by this complex of ideas occurred in the Lausanne region of Switzerland. Between 1438 and 1498, inquisitors operating in this region placed twelve men and seven women on trial for witchcraft conceived in this novel sense. As Kieckhefer has noted, the trials were remarkably consistent, with each confession following a "standard script" that displayed a common narrative structure and provided similar incidental details. Inquisitors were, he argued, interrogating suspects in such a manner that the confessions that they elicited would conform to their expectations. These new ideas about witches and their prosecution gradually began to spread. For example, in 1462 an inquisitor deputed by the inquisitors of Lausanne tried four women in Savoy. These proceedings followed those of Lausanne so closely that Kieckhefer concluded, "Clearly an interrogatory from Lausanne was being used in these trials."[44] The new complex of ideas had reached Italy by the middle of the fifteenth century, where, as Kieckhefer has shown, it became fused with local witch mythologies. In around 1460, two Italian Dominicans, Girolamo Visconti and Giordano da Bergamo, wrote texts in which they conflated the Italian myth of the *strega* with the ideas of witchcraft developed in western Switzerland. In this manner they began to collapse the differences between these local traditions and began to create a new unified mythology that would form the basis for the later fully elaborated concept of witchcraft.[45]

During this period, these new ideas were taken up by some inquisitors. In 1455, a Dominican inquisitor wrote that residents of the town of Edola in the Valcamonica were refusing sacraments, worshipping the devil, and even sacrificing children. By the 1460s, several inquisitors, mainly drawn from the Observant Dominican Congregation of Lombardy, had begun to conduct trials informed by this novel concept of witchcraft. They also subsequently featured in several investigations in the Valcamonica. This indicates that inquisitors operating in particular regions of Italy were pursuing a campaign against forms of magic and witchcraft that would have been unknown to their forebears. It is, however, important to stress that the new ideas did not displace older concerns. Some of the trials conducted in this period continued to be concerned with sorcery. For instance, in 1465 the inquisitor of Bologna convicted the Servite friar Giovanni di Niccolò Faelli da Verona of practicing ritual magic.[46]

Using lists of trials provided by Andrea Del Col and Michael Tavuzzi, it is possible to gauge the impact of these developments on the inquisition of magic in Italy. There was a marked rise in the overall rate of prosecution. During the period 1400–49, there had been nineteen trials for magic in all courts. Between 1450 and 1490, this figure rose to forty-eight individual trials. The most striking rise was in the number of trials conducted by inquisitors, which increased from just five in the earlier period to thirty-five. It is difficult to assess how these developments affected the number of individuals being investigated and executed. Although we know that in the earlier period, thirty-two people were investigated and eleven executed in total, a lack of contemporary documentation makes it difficult to accurately assess the scale of investigations in the later period. Using a sixteenth-century source, Tavuzzi suggested that between 1460 and 1482 the inquisitor of Vercelli, Ivrea, Novara, and Como, Niccolò Constantini da Biella, may have executed as many as three hundred witches, but he could not provide trial documents to support this figure. It is equally difficult to verify information for specific investigations. In 1485, Lorenzo Soleri da Sant Agata, the inquisitor of Vercelli, Ivrea, Novara, and Como, oversaw a series of witchcraft trials in Bormio. As Tavuzzi noted, the author of the *Malleus maleficarum* suggested that it may have led to the trial and execution of forty people, but Del Col was able to find further evidence for only five executions. To take another example, after trials for witchcraft the inquisitor of Brescia condemned an unknown number of women to death in the Valcamonica between 1485 and 1487.[47]

The existing evidence makes it possible to suggest differing assessments of the extent of the inquisition of magic in the period 1450–90. They range from a relatively limited sequence of trials similar in scale to those that took place in Lausanne, which involved fewer than one hundred individuals being tried and executed, to a far more traumatic series of massacres potentially involving several hundred victims. Although it is not possible to arrive at a precise assessment of the scale of trials for magic and witchcraft in the late fifteenth century, we can draw some tentative conclusions. It is clear that inquisitors prosecuted cases of magic more frequently during the second half of the fifteenth century. Furthermore, the overall number of death sentences that they imposed increased appreciably, but it remains uncertain to what extent. It is also evident that some individual trials involved the investigation of unprecedented numbers of suspects. While trials conducted earlier in the century typically involved accusations being leveled at one or two people, one that took place in Peveragno in 1489 resulted in the execution of fifteen women. It is, however, important to stress that these large-scale investigations remained relatively unusual and highly localized events. They occurred largely in northern Italy and were primarily conducted by inquisitors drawn from the Observant Dominican Congregation of Lombardy.[48]

The inquisitors' campaign against magic and witchcraft did not receive universal assent. In 1487, the Venetian authorities prevented the execution of the sentences recently imposed on an unknown number of women from the Valcamonica by the inquisitor of Brescia. Members of the clergy also objected to the new concept of witchcraft. Bernadette Paton has examined the sermons of three Sienese friars, one Franciscan and two Dominican, that were delivered between circa 1460 and circa 1480. She showed that they were broadly concerned by superstition—conceived as the corruption of religious practice—and specifically criticized the practice of magic. On Aquinas's authority, the Dominican preachers condemned the activities of learned necromancers and diviners, and each friar expressed concern for the practice of maleficiant magic among the lower orders. Nevertheless, each of these friars explicitly rejected the new idea of witchcraft advocated by some of their confrères, and at least one of them sought to disprove belief not only in the power of *streghe* but also in their very existence. As we shall see, skepticism about and indeed resistance to the new model of witchcraft was a significant factor driving the production of the most notorious witch-hunting manual of the age, the *Malleus maleficarum*.[49]

The Malleus maleficarum

In 1484, Innocent VIII promulgated the bull *Summis desiderantes*. It opened by affirming the existence of witches who used demonic magic to inflict maleficia. It also recorded the fact that the work of two inquisitors engaged in a campaign against witches in Upper Germany, the Conventual Dominican Jacobus Institoris (1430–1505) and the Observant Dominican Heinrich Sprenger (1436/8–95), had been hampered by local religious and secular authorities. The bull continued by suggesting that local authorities had challenged the inquisitors' right to prosecute witchcraft on both intellectual and jurisdictional grounds. As we have seen, objections to the inquisition of magic were not new, but the conciliar crisis lent them a new potency because they raised contentious questions about the relationship between the papacy and the episcopate. The bull rebutted a technical, jurisdictional claim that Institoris and Sprenger lacked the authority to operate in specific regions and invoked apostolic authority to reaffirm the inquisitors' privileges and jurisdictional right to prosecute heresy. It concluded by commanding the bishop of Strasbourg, Albrecht of Bavaria (1478–1506), to ensure that their work was not impeded.[50]

We learn more about the secular and clerical authorities' objections to Institoris and Sprenger's activities from the text of their notorious witch-hunting manual, the *Malleus maleficarum,* which was first published in Speyer in 1487. It was ostensibly penned by both Institoris and Sprenger, although historians often question the extent of the latter's role in its production.[51] The published work was prefaced by two documents: the *Summis desiderantes* and an Approbation—that is, a document drawn up under public oath at Institoris's instigation.[52] The Approbation lamented that in their sermons members of the local clergy had claimed either that "sorceresses do not exist or that they are unable by any working to bring about an effect resulting in harm to creatures."[53] The Approbation continued that although Institoris and Sprenger wrote the *Malleus* to provide the tools necessary to eradicate witchcraft rather than educate ignorant clergymen, it was nevertheless necessary to disprove their erroneous views.

The Approbation concluded by expressing one further concern: that those clerics who had denied the reality of witches might—through their ignorance of scripture—suppose that the *Malleus* was poorly supported by the views of the doctors of the Church. These comments revealed that Institoris and Sprenger's dispute with their fellow clergymen concerned more than the reality of magic. It raised fundamental questions about the correct definition of Christian truth and the means by which it should be established, by whom, and on what authority. The Approbation recorded that the *Malleus* had been submitted to theologians from the University of Cologne "for examination and comparison against scripture," and it reproduced a series of signatures testifying to their support for the claims that Institoris and Sprenger had made in the text. By soliciting these attestations, Institoris mobilized the authority of the theological faculty of Cologne to assert that the ideas expressed in the *Malleus* constituted the orthodox definition of magic and correctly detailed the inquisitor's jurisdiction over these matters.[54]

Institoris and Sprenger had powerfully asserted the orthodoxy of their conception of magic and superstition, but what ideas did they defend? The *Malleus* was divided into three parts. The first continued the work of the Approbation by invoking a series of arguments drawn from theology, philosophy, and canon law, but above all from the works of Aquinas, to prove the reality of witchcraft. The second, which made extensive use of Nider's *Formicarius*, described the actions of witches and contained the most detailed description of the new stereotype of the witch. They also spoke approvingly of the actions of the inquisitor of Como, whom they claimed had burned forty witches in Bormio.[55] The final part, informed by Eymerich's *Directorium inquisitorum*, described the technical legal processes involved in the work of inquisition. Although Institoris and Sprenger used Eymerich's handbook to discuss the process of inquisition, they did not draw on his ideas of heretical magic or his justification for the prosecution of its practitioners in the first part of the *Malleus*. Historians have justly stressed the crucial role that this text played in articulating and disseminating the new, elaborated theory of witchcraft. Less attention has been paid to the continuities between this work and earlier conceptions of magic and superstition and, in particular, to the significance and nature of its authors' engagement with Aquinas's thought contained in the first part.[56]

In the opening question of the first part of the *Malleus*, Institoris and Sprenger asserted that their work represented true Christian belief. Developing this theme, they considered whether it was a Catholic proposition to defend the existence of sorcerers and whether arguing the opposite was heretical. Those who cast doubt on the reality of sorcery had, they noted, made one of three heretical errors, previously identified by Aquinas in his *Scriptum super sententiis*. The first was to maintain that there was no sorcery in the world and that all effects supposedly caused by sorcerers could be ascribed to natural causes. Aquinas, they continued, had rejected this opinion by arguing that scripture showed that God had allowed demons to exercise control over corporeal objects and the human imagination. Denying that demons could in principle be involved in the creation of magical effects therefore amounted to a conscious rejection of key principles of the faith. Consequently, if people maintaining such beliefs were baptized, they would be liable to censure for heresy. In this instance, the targets of Institoris and Sprenger's arguments were the members of the secular clergy who preached against the inquisition of sorcery, but their reasoning applied equally well to any philosopher—or indeed any other Christian—who might seek to deny the reality of magic by solely invoking natural explanations for experienced phenomena.[57]

Institoris and Sprenger then turned to consider two further skeptical positions that Aquinas had identified. The first was the idea that sorcerers only imagined that they cooperated with demons to achieve their effects, and the second was the belief that although sorcerers genuinely cooperated with demons, the effects they achieved were imaginary. Neither proposition necessarily denied the reality of demons, but they did threaten to diminish sorcerers' powers. They were rooted in two errors that arose, Institoris and Sprenger claimed, from the misinterpretation of Catholic tradition, specifically two sections of the canon *Episcopi*. First, critics of witchcraft had misunderstood a passage from the canon that argued that the women who claimed to ride at night with Diana had been deceived by demons. Skeptics had erroneously assumed that this argument applied equally to all cases of magic and not simply to the Riders of Diana. Second, the canon suggested that it was impossible for any power other than God to change a creature, whether for the better or worse, or to turn it into another creature altogether. Taking the canon's claim that creatures could not be changed for the worse to also mean that they could not be affected by *maleficia*, the skeptics had claimed that sorcery produced no real effects and that any effects that it did produce existed only in the victim's mind.[58]

These arguments presented Institoris and Sprenger with a complex challenge. They were reluctant to dismiss the authority of the canon, but equally they could not let its misapplication stand. They cited passages of scripture, the writings of Augustine and Aquinas, and civil and canon law to reject the proposition that witchcraft and its effects existed only in the imagination before providing what they termed "a healthy understanding of the Canon." Witchcraft skeptics, they argued, had understood it in "simple terms" that were "contrary to the sense of Scripture and the determination of the Doctors." Indeed, they continued, the canon that immediately followed *Episcopi* in Gratian's *Decretals, Nec mirum,* gave a clearer account of the truth. This canon was derived from Isidore of Seville's *Etymologiae*, but Gratian, like previous canonists, had attributed it to Augustine. It acknowledged the possibility that some sorcerers could create real, imperfect creatures. They supported this idea by invoking Augustine's argument—also cited by Nider—that God had planted in the natural order certain seeds that could be manipulated by men and demons. According to Institoris and Sprenger, these passages showed that, with God's permission, demons could indeed work real effects by art, including changing or creating simple creatures. These arguments, they claimed, showed that when read correctly, the canon *Episcopi* did not deny that demons could create real effects. Institoris and Sprenger thus simultaneously maintained the authority of the canon while wresting control of its interpretation.[59]

In subsequent questions, Institoris and Sprenger began to deploy material derived from Aquinas's *Summa theologica* to address further doubts concerning the reality of witchcraft. In question 2, they considered whether in order to work maleficiant magic it was necessary for demons to co-operate with humans and whether humans could perform maleficiant magic unassisted. Institoris and Sprenger's broad conclusion was that neither demons nor sorcerers could achieve anything without the other.[60] To make their case, they had to discount the possibility that humans could legitimately create magical effects through artifice. One of the examples they raised was the act of producing images to capture celestial influences—that is, the art of producing astrological talismans. Specifically citing the *Summa theologica*, they accepted Aquinas's argument that it was always unlawful to make such images and that it therefore followed that by performing this act an operator necessarily entered into a tacit pact with a demon. Indeed, they added, demons incited humans to create such images in order to serve their own true purpose: causing offense to God. Although Institoris and Sprenger were here using Aquinas's ideas to make a specific point about the nature of witchcraft, they also had the effect of drawing attention to superstitious magical practices that had previously been marginal to inquisitorial practice. Moreover, since they stressed that Aquinas had rejected these practices even though they had no necessary ritual element, their discussion of the legitimacy of a particular art—like that of Nider before them—focused on the question of whether that art was naturally possible rather than whether the operator had explicitly invoked a demon.[61]

Institoris and Sprenger's discussion also touched on the legitimacy of astrology. In question 5, they considered whether it was acceptable for a Christian to argue that the increased levels of witchcraft they had encountered in the late fifteenth century were caused by the movements of the stars. Their discussion once again revealed a strong Thomist influence. They began by observing that the stars' influence over an individual's character could be understood in two ways: "Either it is a necessary and sufficient cause or a conditional one that gives a tendency." The former conception of astral influence was, they declared, "not only false but heretical." Elaborating their position, they explained that if the stars were the necessary and sufficient cause of human actions, then human choice would be nullified. Consequently, individuals could not be held to account for their actions, including on any occasions on which they practiced diabolic magic.[62]

Institoris and Sprenger's unequivocal rejection of astral determinism was unexceptional and consistent with the thought of virtually all Christian authorities, including both Augustine and Aquinas. They nevertheless noted that "if, on the other hand, one says that the character of humans is varied by the disposition of the constellations in a conditional way that creates a tendency, then this can be true, since it is contrary neither to reason nor to the faith." They then proceeded to defend the general principle that the stars could exert an influence over the human body and that by these means they could also affect (but not determine) human character and behaviors. Citing the authority of a medical text—Gilbert of Poitiers's *Liber sextus*—they maintained that since the stars could affect the humors, it followed that they could affect character and behavior, albeit indirectly and from a distance. Consequently, although the stars' influence on the humors was real, it was less pronounced than that of terrestrial factors such as the weather or climate. Having conceded the stars' potential influence, they also endorsed the belief—defended by both Aquinas and perhaps Albertus Magnus—that astrologers could use their knowledge of stars' dispositive effects to offer uncertain predictions about future events, including human actions.[63]

Although Institoris and Sprenger's concept of astrological influence was consistent with the views of Aquinas and Albertus Magnus, curiously they did not directly cite their works in this section of the *Malleus* but instead chose to draw on a passage from *De civitate Dei*, in which Augustine had considered whether the stars could affect the health of twins. In this passage, Augustine had contrasted the views of an astrologer, who suggested that their health had been determined by the stars and could therefore be predicted by astrology, with those of a physician, who suggested that it was influenced by environmental factors. This was a surprising choice of text to use to defend predictive astrology, for Augustine had used this story to reject astrological ideas and practices tout court. Seemingly oblivious to this fact, Institoris and Sprenger suggested that he had merely preferred the physician's explanation of the twins' health because it was "more specific and direct." According to their interpretation of this passage, Augustine had believed that the stars could exert an influence over bodies, but he had maintained that their influences were to be assigned a relatively low place in a hierarchy of causes acting on the humors. This allowed them to conclude, seemingly on Augustine's authority, that although the stars could in principle dispose some individuals toward practicing witchcraft, they could not determine this behavior.[64]

It is difficult to establish why Institoris and Sprenger defended their conception of astrology by using this passage from Augustine rather than, for example, the arguments developed in the *Summa theologica*. Their decision may have reflected the church fathers' enhanced authority in this period.[65] The esteem in which Augustine was held does not, however, explain why they chose to cite a passage from his works that contradicted their views. They may have simply been mistaken in their interpretation of his argument. It is also possible that they read his works through a

Thomist lens, which in turn led them to reshape the arguments that they contained and use them to buttress their existing opinions. Whatever the reason, Institoris and Sprenger had subtly misrepresented Augustine's position. Perhaps more significantly, and whether consciously or not, their interpretation of his work also served to efface the differences between the thought of Augustine and Aquinas regarding the acceptability of astrological prediction. As we shall see, the tensions between the thought of these two authorities would become apparent to later commentators.

Marsilio Ficino and Magic

The revival of interest in Aquinas's writings on superstition contained in the *Summa theologica*—evident in both the *Praeceptorum* and the *Malleus*—gave rise to a renewed emphasis on examining the natural possibility of specific arts such as the creation of astrological talismans. The case of Florence offers a means to consider how these developments affected the discussion and practice of magic among learned elites. Under the patronage of the Medici family, a circle of scholars began to gather around the physician and philosopher Marsilio Ficino (1433–99) during the mid-fifteenth century. Captivated by newly discovered Platonic and Neoplatonic texts, Ficino and his followers developed a series of ideas that proved highly controversial. They included formulating novel ways to conceptualize Christian belief and its relationship to pagan philosophy. The Platonic tradition, they claimed, contained divine truths that supported Christian belief, and it was therefore the most appropriate philosophical tradition to use to defend the faith. This formulation of Christian orthodoxy not only challenged Scholastic syntheses of knowledge, but its emphasis on the divine origins of philosophy also had the potential to create a new form of elite religious experience, which threatened to elevate human reason over faith.[66] These latent tensions were clearly exposed in 1486 when one of Ficino's pupils, Count Giovanni Pico della Mirandola (1452–94), proposed to debate nine hundred theses in Rome. Some of the proposed ideas—such as the thesis that there was no better defense of Christianity than the Kabbalah—caused such an uproar that Pico was placed on trial for heresy.[67]

Exploration of the Platonist philosophical tradition also encouraged Ficino to read the works of authors such as Plotinus (c. 205–270), who had described various forms of astrological magic. These influences can be seen in Ficino's work *De vita triplici*. Written over the course of the 1480s, it initially circulated in manuscript before being published in 1489. The work offered scholars advice on how best to manage their health and, in particular, how to ward off the effects of melancholy, an affliction to which they were particularly prone. The first book showed its reader how to construct a regimen suitable for those engaged in the scholarly life, advising on diet, routine, cleanliness, and suggested suitable levels of exercise, rest, and coitus. The second provided advice concerning how best to prolong life. Although many of the remedies were rooted in mundane physick, the work was permeated by an interest in the effects of the stars. It suggested how astral influences could be harnessed or, when necessary, ameliorated. At various points in the first two books, Ficino discussed how individuals could partake of the various qualities that the stars had imparted to certain plants, stones, or metals. He also provided recipes, medicines, and remedies and described how they should be prepared at astrologically propitious moments. This broader theme continued into the third book, "On Making Your Life Agree with the Heavens," in which Ficino explained how individual planets affected the body and the measures that an individual should take to enhance or moderate their effects.[68]

When discussing these various astrological arts, Ficino gave no indication that he or anyone else might consider them heterodox. This approach contrasted sharply with the defensive tone he adopted in "On Making Your Life Agree with the Heavens" when he began to discuss astrological talismans. First, Ficino attached an apologia to *De vita triplici* before its publication, in which he argued that if any reader did not approve of talismans they could always ignore these passages, before adding—perhaps disingenuously—that, in any case, he did "not so much approve of them as describe them." Ficino also adopted a second and more complex defense of the use of talismans, demonstrating that they could, in fact, be considered perfectly orthodox. He began by drawing distinctions between different forms of image magic. The ancients, he observed, had identified several. The examples he cited included making images in the hour of Saturn to cure kidney stones or afflictions of the genitals. Christians such as Pietro d'Abano, he noted, had also approved the use of such images. Ficino then proceeded to discuss images that could work other types of effect, such as bringing happiness or calamity to a particular house or country. He continued by noting that although he did not personally believe that images could achieve these effects, on Plotinus's authority some astrologers—Porphyry, for example—had believed that they could. He therefore maintained that while some images might produce no effects and others might do so only by appealing to demons, it was nevertheless possible that still others might produce genuine, natural effects. The problem therefore was determining which images were legitimate. In the *Speculum astronomiae*, he continued, Albertus Magnus had taught the correct way to distinguish between licit and illicit images. The great theologian and philosopher, he wrote, had accepted that well-made images acquired their powers from a "figure of heaven." He noted that Albertus had also described some images that could be used to work malign effects, before adding that he had discretely chosen not to repeat this information. Although Ficino did not explicitly endorse any of the ideas that he described, he had nevertheless demonstrated the existence of a repertoire of arguments that could be used to construct a defense of the orthodoxy of talismans.[69]

The most significant challenge facing Ficino's attempt to argue that astrological magic could be considered orthodox was the objections that Aquinas had raised in the *Summa theologica*. Ficino did not directly acknowledge these arguments but instead noted that "Thomas Aquinas, our leader in theology, was more afraid of these things and attributed less power to images [than Albertus Magnus or Pietro d'Abano]." Ficino maintained that Aquinas had believed that images could acquire no more power from the heavens than could plants or other natural things. Ficino then invoked Aquinas's discussion of talismans in the *Summa contra gentiles*, which included what Rutkin has termed his "loophole," concluding that Aquinas "does not laugh at images themselves, however, but only at certain signs which are to join them to daemons." In other words, he suggested that Aquinas—in the *Summa contra gentiles* at least—had not condemned the use of images per se but only the use of specifically necromantic images. This selective reading of Aquinas's work enabled Ficino to suggest that his views were similar to those of Albertus, implying that he had in fact accepted—or had at least not condemned—the production and use of astrological talismans. Finishing his discussion, Ficino once more reminded his readers that, in his opinion, it was in any case wiser to restrict oneself to making medicines and to avoid images altogether.[70] Ficino's self-consciously cautious treatment of this subject may have been a tacit recognition of the fact that Aquinas's *Summa theologica* was acquiring a new prominence in discussions of the boundaries of orthodox magic.[71]

Savonarola and Astrology

By the late fifteenth century, Aquinas's *Summa theologica* had become an important resource for defining the boundaries of magic within the mendicant orders. The writings of the Observant Dominican Girolamo Savonarola (1452–98) suggest, however, that his views on superstition were neither interpreted nor used in a consistent manner. Born in Ferrara, Savonarola entered the Dominican convent in Bologna in 1475. During a chapter meeting of the Lombard Congregation, held in Ferrara in 1482, he met Pico, then a student at the University of Padua. The two men renewed their acquaintance in the mid-1480s when Savonarola was appointed lector at the Convent of St. Marco in Florence. He subsequently resumed this position in 1490 with Medici support.[72] Although brought to Florence by the Medici, Savonarola soon began to preach sermons infused with apocalyptic imagery, which criticized the abuses of tyrants, condemned clerical corruption, and called for reform and repentance. From 1493, he began to prophesy the descent of a new Cyrus from the North who would initiate a wide-ranging reform. These predictions appeared to come to pass in 1494 when the French king, Charles VIII (1470–98), descended into the Italian peninsula at the head of a large army in pursuit of a territorial claim to the kingdom of Naples. The French onslaught destroyed the relative stability that had been secured by the Italian League since the 1450s. The Medici, who were allied with the papacy against France, were driven from Florence. In their absence, a new political order guided by Savonarola began to reshape the city. Under his influence, the Florentines sought to rid their city of vice, enacting laws to control sodomy and drunkenness, and in 1495 they established the city's first *mons pietatis*.[73]

Savonarola's work *Tractato contra li astrologi* (1495) reveals that he was also deeply concerned by the threat of divination and magic. In the work's introduction, he recalled that he had often contemplated the threat that astrology posed to Christianity, especially since "the men who are established in high positions are involved in this error, from which they teach the *inferiori* to err."[74] He also noted that he had been inspired to write his own work by reading Pico's *Disputationes adversus astrologiam divinatricem*, a rigorous critique of contemporary astrology, which had been published posthumously in 1494.[75] Savonarola recalled that while reading it he had been saddened that his associate had died so young but nonetheless pleased that he had provided a timely critique of astrology suitable for use "in our times in which almost the whole world is involved in this pestilential fallacy."[76] Savonarola's rhetoric, although perhaps hyperbolic, was rooted in truth. At the end of the fifteenth century, increased technical predictive capabilities and the advent of print had combined with widespread political disorder to create unprecedented demand for knowledge of the future across all orders of Italian society.[77] Savonarola drew on Pico's work to compose his anti-astrological tract but noted that since he intended his work to be read by the "common people" (*vulgari*), he would neither translate the *Disputationes* as a whole nor even attempt to relate its entire content. His aim was merely to show his audience that divinatory astrology was false, superstitious, and contrary to the faith.[78]

The *Tractato* was a relatively short work, consisting of no more than twenty-seven folios. It was divided into three parts, in which Savonarola sought to demonstrate that judicial astrology was false according to theology, false according to philosophy, and, finally, incoherent in itself. In the first chapter of the first treatise of the *Tractato*, Savonarola addressed the question of the status of divinatory astrology in scripture. He began by offering a definition of the nature of divination, which indicated the fact that he had been influenced by the Thomist revival of the fifteenth century. It read:

> They are not called diviners, those who foretell things, which ordinarily proceed from natural causes, either always or almost always, because this is granted to man and it is a human thing. But those who, without divine illumination, presume to foretell future things, which do not have a natural cause, that is, those things that indifferently can either be or not be, and, above all, those things that appertain to the free will, are properly called diviners in Scripture. This is because they attempt to do that which belongs to God alone: and Scripture speaks against this in many places.[79]

Savonarola's definition of divination echoed that of Aquinas in two important respects. First, he did not condemn all who sought to predict future events but only condemned those who sought to foretell things known only by God. Second, Savonarola invoked a characteristically Thomist justification to identify legitimate forms of prediction. Echoing the arguments that Aquinas had formulated in the *Summa theologica*, he maintained that humans could foreknow events with natural causes. Savonarola's approach was therefore underpinned by the reorientation of learned conceptions of superstition that had occurred within the Dominican order during the fifteenth century. Like Nider and Institoris and Sprenger before him, he focused his discussion of the legitimacy of a particular operative art on the question of whether it was naturally possible rather than whether it required the operator to make an explicit compact with a demon. This allowed him to assert the theoretical legitimacy of certain forms of explicitly natural prediction while condemning arts that he did not consider natural.

In his definition of divination, Savonarola had indicated that there may be some acceptable forms of prognostication, including forms of predictive astrology. He did not, however, devote any space in the *Tractato* to discussing practices that he considered legitimate. Instead, he focused on presenting arguments against those activities that he rejected. Opening the subsequent chapter, "*Come è damnata dalli Sacri Theologi*" ("How [Divinatory Astrology] Is Condemned by the Holy Theologians"), Savonarola noted that the theologians of the Catholic Church—meaning here the church fathers and later theologians—were united in their condemnation of divinatory astrology. He began by observing that in *De genesi ad litteram*, Augustine had warned that astrologers occasionally did predict true things, but he maintained that they did so only through the assistance of demons. Consequently, Christians should shun the company of astrologers lest they too become ensnared by demons. Savonarola then discussed passages from *De doctrina christiana*, in which Augustine had argued that although it was possible for humans to accurately track the movements of the celestial bodies, any attempt to use this knowledge to predict the future habits of any individual was false. Savonarola briefly noted that Augustine had used the case of Esau and Jacob to reject any such claim and to counter the idea that differences in time of birth might explain any disparities in their natures. To bolster his case, Savonarola presented further arguments taken from Jerome, Chrysostom, and Origen, but each of their views was dispatched in the course of no more than a few sentences. The last, and the only nonpatristic, authority that he cited was Aquinas.[80]

Savonarola remarked that he had not cited the opinion of every authority who had condemned astrology, but he concluded that "without discrepancy all agree in [condemning divinatory astrology]."[81] Superficially, Savonarola's account of his authorities' views was true. After all, it would be surprising to find a theologian who had defended deterministic forms of astrology. Yet the various authorities that he cited held divergent opinions about the correct definition of "divinatory astrology." Savonarola tended to downplay these differences. When citing Aquinas, for example, he noted that "Saint Thomas also in the *secunda secundae* q. 96 a. 5, and in many other places says that those who, by consideration of the stars, seek to know future things that occur by chance or by accident, and especially the future operations of humans, are foolish and superstitious."[82] Summarized in this fashion, Aquinas's thought appeared to condemn any attempt to predict future contingent events, especially human actions. Nevertheless, in this section of the *Summa*, Aquinas had in fact also endorsed the doctrine of inclination, which maintained that it was possible for astrologers to offer uncertain predictions about future human actions. Although Savonarola did not explicitly reject these ideas, unlike his contemporaries Institoris and Sprenger, he did not endorse them. Inspired by Pico and perhaps also by Augustine, Savonarola seems to have developed a more skeptical approach to Aquinas's work than his Dominican peers. In effect, he appears to have constructed a limited form of "Thomist" astrology that did not admit the idea of inclination.[83]

Later in his treatise, Savonarola considered another of astrology's potential dangers. In chapter 4, he stated that canon law condemns astrology "because it is very harmful to the Christian religion, attributing the mysteries of grace to the heavens, and making the prophets astrologers. So great is the temerity of the astrologers, that they want to attribute to the heavens also the miracles and the martyrdoms of the martyrs that are above any force of nature."[84] Savonarola did not specifically identify the targets of his criticism, but the views he ascribed to these "astrologers" closely echoed those of the philosophers of the universities who adopted a "naturalist" interpretation of the cosmos—that is, those philosophers who sought to interpret Aristotelian philosophy in as accurate a manner as possible, an approach that entailed the elimination of either preternatural or supernatural causation. While contested by many clerics, within the Italian universities the exposition of Aristotle's ideas in this manner remained insulated from ecclesiastical censure. Arts masters continued to use the approach to discussing philosophical ideas developed in thirteenth-century Paris by the Radical Aristotelians—that is, the so-called Averroist approach or "doctrine of double truth." Nevertheless, the radical ends to which they were turned in Savonarola's account—explicit denial of miracles or the ascription of the acts of the martyrs to the movements of the stars—was highly controversial and could not be readily justified by appeal to double truth. While Aquinas had sought to demonstrate the errors in the Radical Aristotelians' arguments, Savonarola bluntly stated, "Certainly one does not dispute with those who say similar things in any way other than with fire."[85]

The Mendicants and Magic at the End of the Fifteenth Century

The pursuit of superstition was central to the mendicants' reform program during the fifteenth century. With the encouragement of the papacy, they had played a leading role in accelerating the prosecution of magic in both inquisitorial courts and secular tribunals. Yet, although prosecutions for magic and witchcraft had increased appreciably between 1450 and 1490, they appeared to decline during the 1490s. Del Col's and Tavuzzi's lists of trials suggest that during this period there were eight trials, of which at least seven were conducted by inquisitors. While the total number of trials conducted during the 1490s was comparable to that in previous decades, each individual trial was on a smaller scale. No more than ten individuals were tried during any individual investigation, and there is evidence for only eleven executions (although the true figure may be higher).[86]

It is also important to note that although a significant number of these trials were connected to allegations of diabolic witchcraft, many were not. In the city of Modena, for example, inquisitors drawn from the Congregation of Lombardy oversaw the trials of six men and two women accused of magic during the 1490s. These trials indicate that despite the growing concern for devil-worshipping witches in some parts of Italy, in other areas inquisitors continued to hear trials for superstition. For example, in 1499 Don Jacopo Pelloni was summoned to appear before the vicar general of the office of the inquisitor to testify about the actions of a certain Bernardina Stadera. He claimed that in his presence she had made paper figurines while uttering words so quietly that he had been unable to understand them. He further alleged that through the ministrations of the nephew of a deceased bishop she had procured chrism and holy oil for the purpose of performing maleficiant magic and that she kept a devil in a glass sphere. In a final revelation, Pelloni pointed to Stadera's connections to—in Kieckhefer's words—a clerical underworld of learned magicians. In her own deposition, Stadera claimed that a Servite friar had lent her a handwritten book that described how to make wax images and have them baptized by a priest, how to curse the mass, and how to perform conjurations. She also revealed that Don Guglielmo, priest and rector of the Church of San Michele in Modena, had shown her a further book of magic and claimed that he could use it to conjure a devil into a phial and ask it questions. Despite this powerful testimony, the inquisitor seemingly did not proceed against Stadera.[87]

If the inquisition of magic was becoming less frequent in the closing years of the fifteenth century, mendicants nevertheless intensified their efforts to restrict the ideas of philosophers who questioned the intellectual bases for belief in the possibility of demonic magic. As we have seen, in the *Malleus maleficarum* Institoris and Jacobus Sprenger had maintained that strict naturalism denied the reality of demonic action and so threatened to undermine belief in the reality of witchcraft, while Savonarola had argued that naturalism was dangerous to orthodox Christianity because it offered a means to explain away miracles. These concerns were also exposed during disputes about the manner in which philosophy was taught and discussed in the universities of Italy. As we have seen, philosophers in the Italian universities continued to use the distinction between true and philosophically necessary arguments to discuss ideas contrary to the faith. By the later fifteenth century, the chief focus of concern tended to be philosopher's discussion of the mortality of the individual soul, but many mendicants also harbored significant concerns about strict readings of Aristotle's *Libri naturales*.[88]

Despite the friars' opposition to the concept of Averroism, it had proved difficult for them to prevent philosophers from discussing ideas in this manner in their manuscripts and books or even in person. To tackle this problem, the mendicants attempted to assert a greater influence within the Italian universities. This approach had been successful in the universities of northern Europe, where the mendicants had been part of the faculty of theology since the mid-thirteenth century. When the Italian universities were initially founded, they did not possess faculties of theology. Consequently, the mendicants' scholarly activities in the great university cities such as Padua and Bologna were restricted to their own *studia*. This situation began to change in the fourteenth century when the papacy granted the University of Padua the right to issue degrees in theology. The mendicants began to teach in the faculty of theology at the University of Padua in 1442, and by 1490 the Franciscan and Dominican orders each occupied a chair in metaphysics and theology. Occupying these positions allowed the mendicants to raise objections to the Averroist style of discussion practiced by many of the philosophers. One of the Franciscans, Antonio Trombetta (1431–1517), solicited the support of the bishop of Padua, Pietro Barozzi (1441–1507). Working with the local inquisitor, the two men issued an edict in 1489, which forbade the public teaching of specific ideas pertaining to the soul. These measures nevertheless failed to prevent philosophers from teaching and discussing ideas in the Averroist manner.[89]

Conclusion

The ecclesiastical history of the fifteenth century was shaped by the conciliar crisis. Its roots lay in earlier disputes about the relationship between the papacy and the episcopate, but the ideas developed by its protagonists pointed the way to the future. They set out a new model for an assertive episcopate to define reform and introduced the idea that the papacy was subject to the council. The papacy responded to the crisis by enacting such measures as redefining the ecclesiological basis of papal supremacy and encouraging the Observant reform movement. Although these measures bolstered papal authority, they also exacerbated tensions with the episcopate. Papal support for the mendicant reform program also had the effect of encouraging the pursuit of superstitious magic, which in turn led to increased prosecutions for these offenses in courts across Europe. These prosecutions were partly driven by a new complex of ideas concerning the figure of the diabolic witch developed within the Dominican order, but the mendicants also continued to preach against and prosecute forms of superstition that would have been familiar to friars active in the fourteenth century. Dominican friars also drove a further intellectual development that was less influential in the short term: the recovery of Thomas Aquinas's writings on superstition and his belief that in order to judge the legitimacy of a contested operative art it was necessary to consider whether it was naturally possible.

5

The Pursuit of Superstition in an Age of Reform, 1500–1517

Introduction

During the fifteenth century, proponents of conciliarism advanced several key ideas. Most notably, they argued that councils should be periodically convoked in order to facilitate reform of the Church, and some advocated the more radical idea that the council enjoyed supremacy over the papacy. After the conclusion of the crisis, the papacy rejected the idea of conciliar reform. Instead, it reasserted the ecclesiological basis of its claims to supremacy over the Church, remodeled the city of Rome, and encouraged the pastoral activities of the mendicant orders. By the beginning of the sixteenth century, it was clear that these measures had not only failed to satisfy many reformers but also served to exacerbate lingering tensions between the papacy and the episcopate. These issues were openly discussed when political circumstances obliged Pope Julius II to convoke the Fifth Lateran Council (1512–17). The events that took place during the period of the council demonstrated the continuation of several distinct programs of reform, which would later shape the debates of the Reformation era. Several of the decrees passed at Lateran V developed ideas of episcopal reform first articulated during the Councils of Constance and Basel. Meanwhile, although the papacy accepted the need for limited reform, it also sought to use the council to protect its authority and privileges. During the period of the council, the pursuit of superstition, magic, and witchcraft reached new heights, but it was not related to any conciliar decrees. This development instead reflected the continued importance of the mendicants' independent program of reform, which rarely influenced the discussions at Lateran V.

The Church and Proposals for Reform in the Early Sixteenth Century

During the fifteenth century, the Christian elite had been divided by the conciliar crisis. At its heart lay a dispute about the respective power and authority of the papacy and the episcopate. The unfolding crisis demonstrated that this issue posed fundamental questions about which individuals or institutions had the power and authority to appoint and dismiss pontiffs and to convoke and dismiss councils, and it raised the contested issue of the autonomy of individual bishops within their dioceses. It also brought to the fore important questions about the nature and purpose of reform and its organization and implementation. The decree *Frequens*, passed at the Council of Constance (1414–18), had obliged the papacy to convoke regular universal councils that would oversee reform. Since the reign of Eugenius IV (1431–47), however, no pope had honored this commitment. After the conclusion of the conciliar crisis, the papacy had instead charted a different course. It had vigorously reasserted the papal supremacy and embarked on a series of monumental building projects designed to reinforce Rome's status as the center of the Catholic world and the pope's position as its high priest. In a further affront to episcopal authority, the papacy had encouraged the expansion of the mendicants' program of pastoral reform through preaching, hearing confession, and the practice of inquisition.[1]

These papal initiatives produced mixed and sometimes unintended results. To finance its building projects, the papacy resorted to a number of expedients including the sale of indulgences and offices. This development exacerbated problems such as simony and absenteeism among the clerical hierarchy, the very abuses that had offended many reformers. Furthermore, it encouraged the further integration of Italy's ecclesiastical and political elites, a development that led them to defend the established system against criticism. The papacy's support for the mendicants' program of pastoral reform also yielded mixed results. Some bishops resented the mendicants' activities, believing that they undermined the bishops' own efforts to engage the population. Others used the quality of the mendicants' pastoral care, which supplemented (and perhaps even replaced) the activities of secular clergy in their diocese, as an excuse for their own nonresidence, thereby exacerbating this problem.[2]

The papacy's strategy of rebuilding its central authority and encouraging reform through agents such as the mendicants attracted the criticism of influential individuals and institutions within the Church and stimulated alternative proposals for reform. The Dutch humanist Desiderius Erasmus (1466–1536), for example, castigated ecclesiastical abuses, criticized the laxity and indeed the very purpose of the regular clergy, and berated what he perceived as Scholasticism's intellectual and spiritual aridity. He also complained that the faith was being reduced to empty ritual at the expense of personal spirituality and inner devotion.[3] Like many humanists, he believed that direct engagement with the word of God could inspire a reform of both the clergy and the laity.[4] Other humanists proposed ways to extend these efforts beyond the learned elites. Some even suggested that members of the laity should have the chance to read God's word in their own language. By the late fifteenth and early sixteenth centuries, vernacular versions of the Psalms, the Gospels, Paul's Epistles, and works of the church fathers began to circulate in increased numbers in Italy.[5]

In Italy, humanism inspired a group of reform-minded young noblemen, who began to gather to discuss spiritual matters in the early sixteenth century. Led by Paolo Giustiniani (1476–1528), the group's members included Vincenzo Querini (1479–1514) and Gasparo Contarini (1483–1542). All were educated at the University of Padua, where they gained a solid training in Scholastic philosophy while also reading widely among the pagan philosophers and modern authors such as Ficino. As Stephen Bowd has stressed, however, it is misleading to label them "Christian humanists," for although they embraced humanist textual practices they remained mistrustful of pagan writings. Like Erasmus, they were critical of the present state of the regular clergy, but they continued to value the form of contemplative life offered by the religious orders. In 1511, Giustiniani entered the hermitage at Camaldoli, where the following year he was joined by Querini. Both subsequently took vows and worked to reform their new order. Deeply affected by his friends' decision to enter the hermitage, Contarini wrestled with the question of whether he too should pursue a similar vocation. During this period, he also began to experience a religious crisis that led him to develop a new set of soteriological beliefs similar to those that Luther would later expound. While not entirely rejecting the role of works in the economy of salvation, Contarini began to emphasize the primacy of faith.[6]

The Venetian circle was linked to a further set of reformers, who met through their shared participation in the Oratory of Divine Love. This confraternity, first established in Genoa in the late fifteenth century, was dedicated to caring for the incurably ill. The Roman branch was established between 1514 and 1517. Its members included two influential reformers. The first, Gaetano Thiene (1480–1547), became a passionate exponent of the charitable values and aims of the oratory. The second, Gian Pietro Carafa (1476–1559), was a member of a wealthy Neapolitan family. Since youth he had been groomed to take high office in the Church. Sympathetic to many reformist ideas, he was also receptive to humanism and had met and corresponded with Erasmus, who had praised his erudition. From an early stage, he possessed a clear vision of reform. On the one hand, he was committed to local reform. As a bishop, he sought to act as a model pastor. On the other, he was a staunch supporter of the papal supremacy. He maintained that corruption in the Church originated within the curia, but he believed that if the papacy could restore discipline and order in the ecclesiastical hierarchy, it would lead to the transformation of the Church as a whole. Despite their contrasting visions, Thiene and Carafa collaborated in founding a new order, the Theatines, in 1524. They conceived the order as a congregation of priests who could inspire the laity through the performance of works of charity.[7]

During the same period, secular and religious leaders made independent efforts to reform the episcopate. For instance, in 1495 Ferdinand of Aragon (1452–1516) and Isabella of Castile (1451–1504) appointed Ximenes Cisneros (1436–1517) to the position of archbishop of Toledo, the most important clerical office in Spain. An Observant Franciscan, Cisneros's activities, like those of Giustiniani and Querini, defy simple categorization. Performing his role with energy and tenacity, he oversaw the reform of the religious orders and the episcopate. He was especially concerned to enhance the bishop's pastoral role by enacting measures designed to limit absenteeism and strengthen local administration.[8] Cisneros was also sympathetic to aspects of humanist reform and played an instrumental role in the establishment of the University of Alcalá, an institution that promoted the study of biblical languages and produced the *Complutensian polyglot*. He nevertheless rejected other proposed humanist methods of reform, notably the idea that scripture should circulate in the vernacular.[9]

The Mendicants, Reform, and Superstition, 1500 to circa 1510

With the encouragement of the papacy and secular authorities, the mendicants had sought to implement a program of reform throughout the fifteenth century. These efforts continued into the sixteenth century and constituted an independent strand of reforming activity, which ran parallel to and sometimes complemented other proposals for reform. At the heart of the mendicant reform program lay an effort to improve pastoral care at a local level while promoting particular forms of behavior and discouraging others. The friars had been particularly concerned by matters such as excessive wealth, Christians' interactions with Jews, and superstition. While superstition referred to any form of corrupted worship, in practice it often meant practicing forms of magic. Since the fourteenth century, the papacy had encouraged the prosecution of cases of magic or divination that involved explicit appeals to demons. This concept of illicit magic remained central to both mendicant preaching and their activities as inquisitors. By the second half of the fifteenth century, however, some mendicants (especially members of the Dominican order) had formulated a more expansive definition of heretical magic. They had been influenced on the one hand by the belief that groups of devil-worshipping witches were seeking to undermine Christendom and on the other by the recovery of Thomas Aquinas's writings on superstition. Consequently, texts such as the *Malleus maleficarum* not only warned of the threat posed by witches but also stressed the danger of ostensibly natural operative arts.[10]

The establishment of this broader definition of heretical magic had the potential to expand inquisitors' pursuit of superstition, but in practice the results were mixed. This was partly due to regional variation. Inquisitorial courts were inconsistently distributed through Europe, and they remained highly localized institutions that operated according to their own practices.[11] To take one important example, in 1478, the papacy permitted Ferdinand of Aragon and Isabella of Castile to create a new centralized inquisition under royal command.[12] The new institution was initially intended to deal with alleged cases of relapsed conversos—that is, Jews who had converted to Christianity and their descendants. From as early as the 1480s, Spanish inquisitors had investigated magic, but these interventions were relatively rare.[13] When prosecuting magic, Spanish tribunals appear to have followed the advice set out in Eymerich's *Directorium*, republished in Barcelona in 1503, which urged inquisitors to concentrate on the prosecution of forms of ritual magic. For example, in 1510, the inquisitor of Saragossa began to investigate a network of learned magicians and uncovered several cases, including that of a priest who possessed a copy of the *Clavicula salomonis*. Accusations of diabolic witchcraft were occasionally heard in Spanish tribunals at the beginning of the century. With the encouragement of the Supreme Council of the Inquisition, such investigations became more frequent between 1525 and 1550, albeit only slightly.[14] The prosecution of magic also remained uncommon in the branches of the Spanish Inquisition established in Sicily and Sardinia in 1487 and 1492, respectively. Of a total of 2,301 accusations lodged with the tribunal in Sicily between 1501 and 1550, only six were concerned with magic or witchcraft.[15]

On the Italian peninsula, the story of the inquisition of magic was more complex. Ritual magic remained a consistent concern for friars and, where the office existed, inquisitors in Italy. From the mid-fifteenth century, inquisitors in certain regions—especially those in the jurisdiction of the Observant Dominican Congregation of Lombardy—had begun to conduct trials for diabolical witchcraft. These prosecutions had been largely confined to the valleys of northern Italy, but by the early sixteenth century, the belief that a sect of diabolical witches posed an imminent danger to Christian society had spread more widely. Tamar Herzig has suggested that the Dominican Domenico Pirri da Gargano, who was inquisitor of Mantua from 1490 to circa 1511, began to investigate cases of diabolical witchcraft in the period 1505–8. His zealous pursuit of witches was, she argued, prompted by his personal connections with Henricus Institoris, the author of the *Malleus maleficarum*.[16] Although the idea of diabolical witchcraft may have been reaching new areas of Italy by the early sixteenth century, the number of trials, which had been in decline since the 1490s, continued to fall. Between 1500 and 1510, there were eleven separate trials in all tribunals, of which only six occurred in inquisitorial courts. The overall number of individuals accused also appears to have been relatively low.[17]

While the number of trials for magic was low in the early sixteenth century, it would also be misleading to suggest that these trials were all prompted by the new idea of diabolic witchcraft. For example, in 1501, the civic magistrates of Perugia conducted a trial against a certain Gniagne Mei di Cibottola. They heard that in April 1481, Gniagne had consorted with "two women, diabolic and maleficiant streghe" in order to perform acts of maleficia. On one occasion, he had conspired with the women to enter the home of a certain Mei Marci de Montelagello, where they drank the blood of an infant child. Gniagne was charged with having been involved in similar crimes on a further four occasions, most recently in March 1500. Although the documents refer to diabolic, maleficiant witches, the activities of the *streghe* recorded in this trial owed little, if anything, to the new ideas being elaborated by some mendicants but were instead consistent with the indigenous mythology of witchcraft that had animated Bernardino of Siena's investigations in Rome in the 1420s. This impression is reaffirmed by the fact that Gniagne was also accused of other offenses, such as attempting to work love magic or cause animosity between a couple by creating enchanted objects and placing them in the houses of his intended victims. Gniagne was found guilty of these crimes and burned at the stake.[18]

The Perugian magistrates' continued belief in the Italian mythology of the *strega* was not the vestige of outmoded concepts and practices but instead reflected the diverse views circulating in late fifteenth- and early sixteenth-century Italy. This plurality of views can also be seen in the *Rosarium sermonorum*, a compendium of preaching materials published in 1496 by the Observant Franciscan Bernardino Busti (1450–1513), which included one sermon that was specifically concerned with superstition and magic. In this sermon, Busti alluded to groups of *streghe*, but he did not subscribe to the idea of a diabolical sect of witches as described in works such as the *Malleus maleficarum*. His idea of the *strega* was instead rooted in the sermons of earlier Franciscans such as Giacomo della Marca and Bernardino of Siena. For instance, he related the stories of Bernardino's involvement in the execution of Finicella and the trial and execution of Santuccia in Perugia in 1445. Busti's decision to include this mythology in his preaching was in some senses as innovative as the Dominican friars' efforts to raise awareness of a diabolic sect of witches. As Paton has shown, in the early fifteenth century many preachers active in Tuscany had dismissed tales of the blood-sucking *strega*. Busti's willingness to use them in his *Rosarium* indicated a continued effort to raise awareness of this mythology of witchcraft, while their appearance in the trial of Gniagne perhaps reflects their wider acceptance.[19]

Busti's *Rosarium* also suggests that his ideas had not been informed by the second broad intellectual change evident in works such as the *Malleus maleficarum*: the recovery of Thomist definitions of superstitious arts. As Fabrizio Conti has shown, Bernardino did indirectly absorb some of Aquinas's insights. For instance, to define superstition he used the writings of his confrère Angelo da Chivasso, who in his *Summa angelica* used the *Summa theologica* to describe idolatry as either offering worship to entities other than God or offering God corrupted forms of worship. Busti did not seem, however, to have drawn directly on the *Summa theologica* to consider the magical arts in detail. In his sermons, he condemned various forms of superstition (including the actions of diviners and necromancers), stressed the dangers of invocation, and warned against producing *breve*—that is, slips of paper marked with words or symbols and used as talismans. Busti's decision to preach against these specific arts may have reflected the forms of magic that he had encountered. He nevertheless condemned these arts on the ground that they involved some form of explicit ritual of invocation. His preaching therefore echoed a conception of superstitious magic stressed by the papacy since the fourteenth century, which, as we have seen, had previously featured in mendicant preaching and recurred frequently during trials for sorcery in the fourteenth and fifteenth centuries.[20]

While Busti condemned arts such as necromancy and the production of *breve*, he did not appear to condemn operative arts that worked by ostensibly natural means. It is unlikely that he had not heard of these arts, for intellectuals certainly continued to discuss them in the late fifteenth and early sixteenth centuries. His silence on these matters may reflect ongoing confusion about their status. As we saw in the last chapter, although the recovery of Thomist thought in texts such as the *Malleus maleficarum* may have reanimated debates over the legitimacy of arts such as the production and use of astrological talismans, Ficino had shown that it was possible to defend their orthodoxy. There were similar ambiguities surrounding the status of arts such as physiognomy and chiromancy. While physiognomy was a well-established subject in European universities and numerous libraries held copies of physiognomical texts, the status of chiromancy was less clear. As we have seen in previous chapters, while Aquinas regarded this art as a form of illicit divination, Nicholas Eymerich used it as an example of a legitimate natural art in his inquisitorial handbook. It seems that Eymerich's opinion continued to predominate in the early sixteenth century, for it certainly remained possible for intellectuals to discuss these arts. For example, Bartolomeo della Rocca Cocles (1467–1504) wrote *Chiromantie et physiognomiae Anastasis*, a work in which he described the practice of physiognomy and chiromancy, provided a number of aphorisms, and explored the connections between these arts and astrology. It was published in 1504 with an approbation by Alessandro Achillini (c. 1461–1512), a professor of medicine and natural philosophy. Although theologians such as Nider and Institoris and Sprenger had recovered Thomist ideas on superstition, it seems that their innovations had not curtailed intellectual discussion of arts that Aquinas had condemned.[21]

The Convocation of the Fifth Lateran Council

Discussion of reform was given a new focus during the pontificate of Julius II. Drawn from the noble Della Rovere family, he embodied many of the contradictions of the Renaissance papacy. Though he owed his position to nepotism and bribery, once elected he did little to enrich his family. As the secular ruler of the papacy, he expended such effort to protect the Papal States that he earned the epithet "the Warrior Pope." A champion of reform within the religious orders, he appointed Giles of Viterbo (1472–1532) prior general of the Augustinian order in 1507, but he nevertheless resisted the idea of a reforming council.[22] His attitude toward a council changed in 1510, when the papacy was drawn into conflict with Louis XII of France. In addition to taking military action, Louis adopted a strategy similar to the one pursued by his predecessor Philip IV in his dispute with Boniface VIII. He persuaded several cardinals to convoke a council in Pisa, with the aim of securing the pope's deposition. These actions were a timely reminder to Julius that the strongest forms of conciliar theory, those that asserted councils' supremacy over the papacy, remained a potent threat to papal authority. Julius responded by forming a military alliance with major European powers such as England and the Holy Roman Empire and by convoking his own council to be held in Rome.[23]

Many members of the episcopate welcomed Julius's decision to convoke a council, believing that it offered them a chance to secure reforms that they desired. Spanish bishops were especially optimistic. Before their departure for Rome, they met in Burgos to agree on a series of proposals to bring to the meeting. Inspired by the reforms spearheaded by Cisneros, they sought to enhance the power of local bishops. As Nelson H. Minnich has observed, they believed that they could also secure a reform of the papacy that would be enforced by the imposition of legal constraints. Their proposals included limiting simony in papal elections and ensuring that cardinals were of a suitable quality and committed to the regular convocation of councils. These proposals were echoed in a series of reforms suggested by the Dominican archbishop of Seville, Diego de Deza (1444–1523). Perhaps reflecting his own experience as a bishop, in addition to remedies for the problems of simony and episcopal nonresidence he proposed limiting the exemptions enjoyed by his fellow friars. The Spanish episcopate's ambitions were supported by the monarchy. At the beginning of 1512, Ferdinand of Aragon drew up a series of instructions for his ambassadors. In this document, he praised councils and set out his belief that they should be called regularly in accordance with the provisions of *Frequens*, but he nevertheless urged Julius to revoke the decrees of Basel and Constance that placed the authority of the council over that of the pope. Echoing the synod of Burgos, he was also strongly critical of simony.[24]

Julius's orthodox council assembled on April 19, 1512. It was well attended, and the fathers were drawn from each of Latin Christendom's major political units.[25] It opened with an address from Giles of Viterbo, who proclaimed the value of councils, noting that their regular convocation was necessary to the health of the Church. With God's aid, he added, they had the power "to root out vice, to arouse virtue, to catch the foxes who in this season swarm to destroy the holy vineyard, and finally to call fallen religion back to its purity, its original splendor, and its own sources."[26] Given the circumstances that led to the council's convocation, it is no surprise that the fathers' first substantive act was to condemn the rival Pisan council and to declare its decrees null and void. Despite this early focus, the fathers also began the work of reform. During Session 5, held on February 16, 1513, they passed decrees designed to control simony. Despite these early successes, the council's future and hope for further reform were thrown into doubt when, just five days later, Julius II died. He was succeeded by Cardinal Giovanni de' Medici (1475–1521), who was elected as Leo X on March 11, 1513. The election of the new pope raised hopes for a reconciliation between the papacy and the French monarchy, thereby removing the primary purpose of the council. Leo nevertheless decided to continue with the council, although he moved to postpone the next session.

The hiatus offered reformers a new opportunity to reflect on what a council could achieve. In summer 1513, the Camaldolese hermits Querini and Giustiniani were in Rome promoting their proposed reforms to their order. During this period, they wrote *Libellus ad Leonem X*, a manifesto for the reform of the Church, which they sent to the pope.[27] In the opening part of the *Libellus*, the hermits set out the scope of their vision. They wanted to see a transformation not simply of Christendom but of the whole world. This was all to be achieved under the leadership of the pope, who, as the successor of Peter, was Christ's representative on Earth. Indicating their ecclesiology, they added that those who sought to equate the pontiff's powers with those of other bishops were in error. To achieve this plan, the hermits detailed how the pope should convert the Jews and the pagans—that is, the indigenous populations of the Americas. Respecting the principle of relative toleration established in canon law, they argued that although the Jews should not be forced to convert, the papacy should encourage this to happen by enacting a series of measures designed to make continued adherence to their faith intolerable. Next, they proposed a crusade against the Muslims, which would have the dual aim of uniting Christians while subjugating the infidels. Turning to their fellow Christians, they suggested that the seven Christian nations should be united under the direction of the papacy.[28]

In the fifth section of their *Libellus*, Querini and Giustiniani proposed a series of reforms that the papacy should undertake, which, they believed, could help revive the Latin West. This section of their text was permeated by medical metaphors, in all likelihood a reference to the fact that their text was addressed to a Medici pope. They offered both a diagnosis of the cause of Christendom's ills and a prescription for their cure. There were, they maintained, two major causes of degeneration in the Christian Church. The first was secular rulers' lust for power, which created conflicts and denied the people peace. To restore peace, they recommended that the pope should seek to divert the secular rulers' energies against the enemies of Christ by calling a crusade. Once this was achieved, the pope should seek to establish justice. In an echo of the rhetoric of Cluniac reform, and no doubt conscious of the concessions wrung from the papacy during the conciliar crisis (such as the Pragmatic Sanction of Bourges), they suggested that the Church should be freed from secular taxes and lay involvement in the process of appointments. Finally, they argued that the pope should deliver wider freedoms for the people, including protecting them from the exactions of the rich and the pernicious effects of usury, whether at the hands of Christians or Jews. If the pope could restore peace and justice in this manner, his actions would eliminate the miseries of temporal life.[29]

The hermits then proceeded to discuss the second cause of Christian corruption: ignorance. They suggested that this evil flowed from the religious—that is, the secular and regular clergy. This was an especially pernicious problem, for if the religious were ignorant, they would be unable to correct the errors of their flock. As a remedy, Querini and Giustiniani advocated improving the clergy's level of education. They insisted that entry to the secular and regular clergy should be made contingent on candidates demonstrating their capacity to read Latin. Achieving such a goal, they maintained, required the wholesale reform of the existing system of clerical education. Their specific proposals demonstrated an ambivalent attitude toward humanism. On the one hand, they made the familiar call for an end to the "dialectical quibbles" of the schools; on the other, they proposed that students should no longer learn Latin and Greek by studying pagan literature but by reading the works of revered Christian authors. Querini and Giustiniani also advocated methods to alleviate the ignorance of the people. They argued that scripture should be translated into the vernacular so that passages might be read aloud in church. While the hermits were prepared to allow the laity to hear the word of God in their own language, they did not appear to countenance the idea of allowing them unmediated access to scripture through the circulation of vernacular Bibles.[30]

Querini and Giustiniani were deeply concerned by the problem of ignorance, but they cautioned that "from a wicked mother comes a daughter who is worse." Ignorance, they wrote, bred superstition, which now thrived throughout Christendom. Although the term *superstition* denoted the corruption of religious practice, like many earlier reformers they were especially concerned by magic and divination. In an important development, they warned of the proliferation of various forms of prognostication such as astrology, chiromancy, hydromancy, pyromancy, and geomancy. Many within the clerical elite, they continued, were simply unaware of the fact that these arts were widely diffused through Italian society. Indeed, they noted, "If Your Holiness should seek out the printed books of the illicit arts, you would find them more numerous, varied, empty and impious than you could have believed or imagined." All such works should, they urged, be seized and burned. They also condemned healing arts, noting with distaste the fact that charms and amulets were sold openly.[31] As we have seen, hitherto the predictive arts mentioned by the hermits had been marginal to inquisitors' concerns, and Eymerich had considered some—notably chiromancy—to be acceptable. The hermits' decision to denounce them may simply reflect the fact that there was widespread demand for knowledge of the future in early sixteenth-century Italy, but I suggest that it instead represents an important change in the elite understanding of magic.[32] The hermits' assessment of these arts reflected the increasing engagement with Aquinas's thought on superstition among theorists of witchcraft and magic such as Nider, Institoris, and Sprenger.

In a further section of the *Libellus*, the hermits discussed healing practices. This passage provides further indications that the hermits had been influenced by the criteria for determining the boundaries between natural and superstitious arts developed by Aquinas in the *Summa theologica*. While discussing healing arts, they argued that "it is certain that whatever does not induce health according to its own nature is superstitious, and is therefore impious and criminal."[33] In other words, it was acceptable to employ remedies that worked by means of a substance's innate qualities but illicit to attempt to use it to produce an effect that it could not naturally cause. Some Christians, they continued, used healing magic and abused religious ceremonies to treat patients. The papacy should therefore forbid anyone without qualification from practicing medicine "and stipulate that nothing may be applied to any sickness that is not consistent with its own nature, and is considered by physicians to have curative powers." This formulation echoed Aquinas's belief that some materials did indeed possess occult healing qualities but that attempts to manipulate them in order to produce further effects was a form of illicit superstition. Once again, the hermits' comments suggest that by the early sixteenth century, at least some of the Catholic elite had altered their understanding of superstition. In addition to explicit forms of invocation, they now emphasized the importance of considering whether a particular art was naturally possible when seeking to determine whether it was superstitious.[34]

Having described the nature and extent of the problem of superstition in contemporary society, Giustiniani and Querini proceeded to offer a highly provocative diagnosis of its cause. First, they accused the episcopate of neglect. They observed that although superstition was widely spread, "not one of our many bishops objects to it."[35] The hermits then leveled a yet more serious charge at the religious. Not only did they fail to address the related problems of ignorance and superstition, but these abominations in fact "originate[d] among religious, and from those that ought to have the cure they spread to the people."[36] In the hermits' opinion, the religious were not simply failing to tackle these threats; they were their primary cause. Their point was almost certainly exaggerated, but it contained a kernel of truth. On the one hand, they pointed to the ambiguous boundaries between Christian religion and magic—for instance, condemning the recitation of prayers falsely ascribed to the saints and the practice of carrying particular images to ward off plague.[37] Such practices were perhaps permitted by ill-educated local priests. Furthermore, as we have seen, priests frequently appeared in trials for magic, whether for supplying consecrated hosts or holy water to magicians or for practicing forms of learned magic themselves.[38]

To eradicate the related scourges of ignorance and superstition from Christian society, the hermits believed that it was necessary for the papacy to initiate and lead a reform of the clergy. They described a model of papal reform that was similar to the one favored by the likes of Carafa but that was, in some respects, at odds with the bishop-led models of reform advocated by the Spanish episcopate. They maintained that the pope must initiate change by rectifying his own behavior. Only by renouncing the failings of his predecessors could a pontiff hope to restore the dignity of the office of the papacy. For Querini and Giustiniani, the personal sanctity of individual pontiffs alone would not suffice to effect the required reform of Christian society. The pope also needed to instruct and—where necessary—correct the errors of the clergy, whether cardinals, bishops, parish priests, or members of the religious orders. The hermits believed that reform of the secular clergy should radiate out from the center. It was therefore necessary for the papacy to begin by reforming the cardinalate. Since the chief cause of sin among the cardinalate was the abundance of benefices, they recommended that cardinals' incomes should be restricted to the benefices of their titular churches and a supplementary pension. The reformed cardinalate would oversee the selection of archbishops and bishops and ensure that they performed their duties adequately. The bishops in turn should oversee the lower clergy. Operating through this hierarchical structure, the papacy could enforce discipline within the whole Church.[39]

The hermits then turned to address the reform of the regular clergy. They demanded the abolition of all remaining conventual branches of the religious orders and stressed that it was the papacy's duty to ensure that every regular observed their order's rule. The hermits also sought to insert the religious orders into their hierarchical system of discipline by making them submit to the scrutiny of the episcopate. They proposed that "all monks, nuns and religious of every kind be subject, as they were not long ago, to the authority of bishops."[40] Although they maintained that bishops should not interfere in elections, they believed that they should be authorized to inspect individual houses to ensure that they were run properly. These changes would, they argued, lead to the reform of society as a whole, for once the papacy had rectified the errors of the clergy, those of the people would also soon be resolved.[41]

Leo X and the Reforms of the Fifth Lateran Council

Lateran V recommenced on December 19, 1513. Session 8, the first of the reconvened council, was dominated by efforts to reconcile the French monarchy and the papacy. The French king's ambassador formally renounced the Pisan council and reaffirmed the monarch's adhesion to the papacy, thus formally bringing the schism to an end.[42] Although the council had fulfilled its initial purpose, Leo declared his intention to allow it to continue. His decision was prompted by a desire to use this opportunity to secure the abrogation of the Pragmatic Sanction of Bourges, which had translated the early decrees of the Council of Basel—including those asserting the principle of the supremacy of councils over pontiffs—into French law. By keeping the council open, he hoped to undermine the juridical basis of strict forms of conciliarism that asserted the council's supremacy over the papacy, but it also meant that he opened the way for weaker forms of conciliarism—that is, those that sought to use councils to secure reform of the Church in its head and members. To secure his objective, Leo might be obliged to agree to accept reforms that strengthened the episcopate vis-à-vis the papacy, thereby restricting papal power while preserving the principle of supremacy.[43]

When the council resumed, consideration of its direction remained in the future. The fathers needed to discuss decrees that had been prepared prior to its suspension. During Session 8, the fathers passed the decree *Apostilici regiminis,* which was designed to address concerns about how philosophy was taught at the Italian universities. It began by noting that "the ancient enemy of the human race, has dared to scatter and multiply in the Lord's field some extremely pernicious errors which have always been rejected by the faithful." These errors included such ideas as the mortality of the individual soul and the argument that all humans shared a single soul, which were propagated by those who "playing the philosopher without due care, assert that this proposition is true at least according to philosophy." This was a reference to the style of philosophical discussion referred to as "Averroism" or "double truth" that continued to be used in the Italian universities, which had been an ongoing source of concern to the mendicant orders. The authors of *Apostilici regiminis* sought to reject this style of discussion by declaring that "since truth cannot contradict truth, we define that every statement contrary to the enlightened truth of the faith is totally false and we strictly forbid teaching otherwise to be permitted." The decree therefore reaffirmed the long-established principle that if a philosopher maintained a position to be true that theology determined to be false, then he was propagating a heresy.[44]

The wording of *Apostilici regiminis* clearly reiterated the nature of heresy, but its practical implications were limited. It could not prevent philosophers from engaging in the public discussion of ideas that contradicted the faith or from articulating them in writing. This was because the university masters had not publicly taught that ideas such as the mortality of the soul were true; if they had, then they would already have been liable to prosecution as heretics. The bull tacitly acknowledged this fact, noting that "it does not suffice occasionally to clip the roots of the brambles, if the ground is not dug deeply so as to check them beginning again to multiply, and if there are not removed their seeds and root causes from which they grow so easily." The decree was referring here to the need not only to prevent individual instances of philosophers publicly expressing ideas contrary to the faith but also to transform the culture of philosophical instruction. The decree required all philosophers, whether teaching in a university or elsewhere, to make clear to their audience when the ideas that they were describing deviated from the faith. Furthermore, it enjoined them to set out the truth established by the Christian religion and "to teach it by convincing arguments, so far as this is possible." This phrasing appeared to acknowledge the fact that although it might not be possible for an individual philosopher to find arguments that supported the faith, such arguments could nevertheless exist. The decree also required philosophers to refute the arguments that were opposed to the faith "since all the solutions are possible."[45]

Apostilici regiminis had several implications for the discussion and practice of learned magic. As the decree suggested, the primary target of this particular measure was ideas concerning the mortality of the individual soul. More broadly, however, it restricted the teaching of philosophers who sought to expound Aristotle's philosophy without reference to the truths of the Christian faith. The decree therefore questioned the validity of Renaissance "naturalism"—that is, the practice of explaining observed phenomena by solely natural means—that was influential in the Italian universities in the later fifteenth century. Many mendicants feared that this form of argument threatened belief not only in the reality of miracles but also in the possibility of magic and witchcraft, and it seems that this specific problem was on the minds of at least some participants in the council. In a *Historia* setting out the dangers of Paduan philosophy that he penned during the council, Giles of Viterbo specifically raised the problem of denying the reality of angels and demons when discussing the natural order.[46] When teaching publicly, philosophers were now obliged to explain to their audience the errors inherent in or arising from a naturalist account of the cosmos. The decree only moderated how these ideas were taught, however. It did not prevent philosophers from discussing them or indeed from publishing books treating them. To take one example, in the years after Lateran V, Pietro Pomponazzi (1462–1525) published such works as *De incantantionibus* and *De fato* in which he appeared to deny the reality of preternatural and supernatural causation within the cosmos.[47]

With Session 8 concluded, the council fathers had dealt with any outstanding business from the previous session of the council. Leo X now invited them to propose further items for discussion. The bishops submitted to the pope a document containing two key demands that were designed to enhance the episcopate's power in relation to the papacy. First, they called for the establishment of a permanent episcopal organization similar to the College of Cardinals in Rome, and second, they called for the abolition of the mendicant orders' privileges. Leo X was initially receptive to the call for an episcopal college, but, following objections from the cardinalate, he began to resist the idea. He did, however, appoint a commission of cardinals to discuss the question of mendicant privileges. A delegation of bishops submitted a list of demands to the commission. Echoing longstanding complaints, they argued that by allowing the friars to absolve any sin (including heresies and reserved cases), preach without episcopal oversight, and collect burial fees, the papacy had undermined bishops' authority in their own dioceses. Fearing for the future of their orders, the Dominican Tommaso de Vio Cajetan (1469–1534) and the Augustinian Giles of Viterbo made submissions of their own.[48]

When the council resumed its business in Session 9, held on May 5, 1514, Leo X attempted to address some of the complaints raised by the bishops by producing a lengthy bull of reform, *Supernae dispositionis*. The reforms carefully balanced the contrasting ambitions of the episcopate and the mendicants. Answering the concerns of the reformers who believed that episcopal residence was central to efforts to improve pastoral care at the diocesan level, the decree addressed the problems of simony and absenteeism by curbing the dispensation of benefices. It also proposed reforms to the head of the Church. It required cardinals to be resident in Rome, limited their households, and urged them not to employ bishops in any "demeaning tasks." Reiterating their role in effecting reforms of Christian society, the decree also exhorted cardinals to visit (or send a deputy to visit) the place of their titular basilica and stressed their responsibility to know which areas were affected by heresy or superstition or where ecclesiastical discipline was lacking.[49] The decree then began to propose measures designed for the reform of Catholic society by defining religious instruction in schools and setting out the penalties for blasphemy. Reaffirming the importance of repentance within the established systems of pastoral discipline, it stated that "no blasphemy can be absolved without a heavy penance imposed by the decision of a strict confessor." Next, the bull addressed the venerable problem of clerical chastity and simony at a local level by reiterating the authority of previously established canons.[50]

Supernae dispositionis implemented a number of measures that were consistent with the reforms demanded by the episcopate, but it also addressed matters that were more closely associated with the reforming agenda pursued by the mendicants during the previous century. It proposed measures to deal with three problems that had been regular features of mendicant preaching: superstition, heresy, and the presence of the Jews in Christian society. While the latter two issues had been the subject of earlier decrees, superstition was rarely if ever mentioned at universal councils. The decree's passage on superstition read,

> Sorcery, by means of enchantments, divinations, superstitions and the invoking of demons, is prohibited by both civil laws and the sanctions of the sacred canons. We rule, decree and ordain that clerics who are found guilty of these things are to be branded with disgrace at the judgment of superiors. If they do not desist, they are to be demoted, forced into a monastery for a period of time that is to be fixed by the will of the superior, and deprived of their benefices and ecclesiastical offices. Lay men and women, however, are to be subject to excommunication and the other penalties of both civil and canon law.[51]

The decree articulated a general aspiration to curb forms of superstition such as sorcery and divination that was consistent with mendicant ambitions. Yet while it reaffirmed the broad principle that those who performed superstitious acts should be punished, this decree did not define the boundaries of orthodoxy or even set out general principles useful for establishing those boundaries. More specifically, the decree's wording showed no indication that it had been influenced by either of the two novel ideas that had been developed by Dominican friars during the previous century—that is, concern for an organized diabolic sect and the renewed focus on analyzing whether specific arts could function by natural means.

The decree continued by declaring that all heretics, those merely tainted by heresy, and all Judaizers "[were] to be totally excluded from the company of Christ's faithful and expelled from any position, especially from the curia, and punished with an appropriate penalty." It therefore once again expressed a general aspiration: that Christian society should be cleansed of these contaminating elements. Again, this ambition was consistent with the mendicants' program of reform, but it had wider implications. The friars' reform efforts had been primarily targeted at the laity. The decree's reference to the importance of expelling heretics and Judaizers from the highest ranks of the clergy underlined its authors' intention to make all members of the Christian community conform to Church law. The ecclesiastical elite were to be no exception; indeed, it was essential that they too were orthodox, because they were exemplars for the whole community.[52]

The council fathers did not comment again on superstition or heresy, but in Session 10, held on May 4, 1515, they passed a further decree regarding the *montes pietatis*. The *montes* were primarily associated with the Franciscans, who established them to provide Christians with a means to avoid usurers and especially Jewish moneylenders. The council, the decree noted, was responding to complaints leveled by some masters of theology and canon and civil law that the system was itself usurious. As Minnich has shown, in the late fifteenth century many of the most important criticisms of the *montes* had been leveled by Dominican and Augustinian friars and answered by Franciscans. This was an important reminder that fault lines within the Church ran not only between the regular and secular clergy but also between the religious orders. While the decree recommended certain reforms, it also defended the principle of the *montes* by unequivocally declaring that "such a type of lending is meritorious and should be praised and approved."[53] It concluded by asserting that anyone who preached or wrote against the schemes in a manner contrary to its provisions were to be subjected to the penalty of excommunication. This measure therefore placed a key tool of mendicant reform, or at least Franciscan reform, beyond reproach.[54]

Mendicant authority was further enhanced during Session 10 when the council fathers passed the decree *Inter sollicitudines* concerning printed books. It noted that "some printers have the boldness to print and sell to the public, in different parts of the world, books—some translated into Latin from Greek, Hebrew, Arabic and Chaldean as well as some issued directly in Latin or a vernacular language—containing errors opposed to the faith as well as pernicious views contrary to the Christian religion." In response to this threat, the council fathers implemented measures that subtly reconfigured the existing system of press censorship. The bull *Inter multiplices* of 1487 had given the bishops—and in Rome the Master of the Sacred Palace, a position traditionally held by a Dominican—responsibility for both licensing texts prior to printing and censoring unlicensed works that were circulating in their dioceses. The decree of 1515 restated these powers, with one important change: from this point forth, local inquisitors—in practice often members of the mendicant orders—would join bishops in the censorial process. This meant that all books, including works of philosophy or learned magic, now needed to be passed by one or the other of these figures.[55]

The council had passed a number of decrees that served to advance causes closely associated with the mendicants, and *Inter sollicitudines* had buttressed their authority in the field of censorship. The episcopate nevertheless continued to press for reforms that would advance their interests. In 1516, the bishops presented a new supplication to Leo X regarding their proposed Congregation of Bishops, which he once again rejected. Despite this setback, during the council's eleventh session, held on December 19, 1516, the council fathers passed a number of decrees designed to strengthen the role of the bishops at a local level. These measures, which later inspired the reforms of Trent, often had the effect of limiting central papal authority in individual dioceses. This point is illustrated by the first decree, which concerned the practice of preaching. While it acknowledged the vital role played by preachers, it stressed that they "ought to adopt only what the people who flock to their sermons will find useful, by means of reflection and practical application, for rooting out vices, praising virtues and saving the souls of the faithful." The council fathers therefore declared that anyone who sought to preach, whether a member of the secular clergy or a mendicant, needed to be examined first by their superior to see if they were suitable to discharge this office. Anyone wishing to preach was now also required to demonstrate to the local bishop that they had undergone such an examination. This meant that although the mendicants could continue to determine which of their brethren were suitable for preaching, the bishop had the right to oversee this practice within his diocese. It also added that preachers were to "refrain from any scandalous detraction of bishops, prelates and other superiors and their state of life."[56]

The decree on preaching struck a delicate balance, allowing mendicants considerable autonomy while reiterating the bishops' right to control the cure of souls within their own diocese. During the same session, the council fathers passed another decree that further defined the respective rights of the episcopate and the mendicants. It opened by acknowledging the good work of both the bishops and the friars, praising their efforts to promote the true faith and maintain Church unity and their work training and saving souls. Gesturing to the ongoing tensions between the secular and regular clergy, the decree suggested that it was appropriate that the council take steps "to contribute to the preservation of peace and quiet among them" by ensuring that "each preserves as far as possible his own jurisdiction." Many of the provisions once again reaffirmed the episcopate's right to control pastoral care in their dioceses. It determined that bishops could, for example, conduct visitations of parish churches held by friars and punish those who failed to adequately perform their duties. The decree also obliged the mendicant superiors to ensure that any friars appointed to hear confession were first presented to the local prelate. If they were accepted, or if any refusal were unjust, the friars were to dispense their duties in conformity with the bull *Omnis utriusque sexus*.[57] Strengthening local bishops' oversight of the cure of souls in their diocese, the decree stated that the ordinary should examine any friar who was due to assume the role of parish priest and that he alone possessed the authority to ordain them. The decree stressed that in all other matters not mentioned, the existing rights of bishops and mendicants were to be maintained. The bishops were not to receive—as Querini and Giustiniani had proposed—visitation rights over religious houses. It concluded by expressing the hope that, with their respective rights and responsibilities having been redefined, the bishops and friars could together "provide a light to all Christ's faithful."[58]

Leo was ultimately rewarded for his stewardship of the council by the fulfillment of his long-held ambition to reset the papacy's relationship with the kingdom of France. The negotiations, which had begun when Leo assumed the papal throne, entered a new stage after the death of Louis XII. In August 1515, his successor, Francis I (1494–1547), led an army deep into Italy, threatening the city of Florence. These actions brought Leo X, in his capacity as ruler of Florence, to the negotiating table, where he signed the Concordat of Bologna. This required him to make significant concessions to the French king, including acknowledging his right to appoint archbishops, bishops, and prelates within his kingdom, although these appointments would remain subject to the confirmation of the papacy. In effect, the French monarchy gained control over the French Church. In return, the papacy was assured of its rights to collect ecclesiastical revenues in France and secured the revocation of the Pragmatic Sanction of Bourges. These agreements were set out in the decree *Pastor aeturnus*. It asserted that, except in situations in which there were two or more rival pontiffs, the pope enjoyed the sole right to summon, transfer, and dissolve councils. Despite the various concessions that he had made to both the episcopate and the French monarchy, Leo hoped that this decree would finally remove the threat posed to the papal supremacy by strict forms of conciliarism.[59]

The Prosecution of Magic during Lateran V

While Lateran V was in session, Italy experienced a notable surge in prosecutions for magic and witchcraft. The total number of trials rose to seventeen, appreciably higher than it had been during the preceding two decades, and the number of people investigated and executed also rose sharply. Many of the largest hunts occurred once more in the valleys of northern Italy. In Como, around thirty people were executed in 1513–15. In the Valcamonica, one hundred witches were burned in 1516–17; a further hunt began in 1518, which would continue for a further three years. Each of these three large-scale witch hunts were conducted by inquisitors drawn from the Congregation of Lombardy, and they were all brought to a close by the intervention of the Venetian secular authorities.[60] There was also a cluster of fifteen trials in Modena, which began in 1517 and ended in 1523. This surge in trials for magic was contemporaneous with Lateran V, but it was not caused by decisions taken at this meeting. As we have seen, the Lateran fathers had addressed the problem of superstition in May 1514, but their decree simply reaffirmed existing canons. The impetus for this increase in rates of prosecution came instead from the mendicant orders.

The large-scale trials in the valleys of northern Italy resulted from a renewed campaign against diabolic witchcraft. Tamar Herzig has argued that at least some of these trials were conducted by inquisitors influenced by personal contact with Henricus Institoris, the author of the *Malleus maleficarum*, who had traveled to northern Italy in the late 1490s and early 1500s. She further suggested that inquisitors' firsthand acquaintance with Institoris's ideas encouraged the diffusion of witch-hunting beyond the valleys of northern Italy to "the bustling urban centers of Emilia, Romagna and Lombardy." She observed that "the growing tendency to discern signs of witchcraft (rather than mere superstition or sorcery) is clearly manifest in the records of trials conducted at Modena in the years 1518–1520, most notably in the trial of Anastasia 'la Frappona.'"[61] It is beyond dispute that some inquisitors active in early sixteenth-century Italy were influenced by the elaborated stereotype of the witch, and Herzig's case for Institoris's influence on the views of a number of individual inquisitors is compelling. However, her argument can offer only a partial explanation for the cluster of trials that occurred in Modena between 1517 and 1523.

Further examination of the documents generated by the Modenese trials shows that the elaborated stereotype of the witch had not displaced concern for older concepts of superstition. Matteo Duni has demonstrated that although these trial documents contain occasional references to rites similar to the "Games" of Oriente discussed in chapter 3, only from the

1530s is it possible to find "trials of women who confess to having been 'witches' in the full sense."[62] Furthermore, the trial documents show that the elaborated stereotype of witchcraft played a less significant role in Anastasia's trial than Herzig intimated. They record that in 1519 Anastasia was interrogated about a variety of offenses including killing her husband by magical means and performing various idolatrous acts. One document, a summary of the final charges leveled against Anastasia, contained a brief reference to the possibility—not explored in the extant interrogations—that she may have been an apostate, who had abandoned the Catholic faith to join a sect that took the devil as their Lord. Although this was indeed a reference to the elaborated idea of witchcraft, these ideas played no role in either Anastasia's trial or her subsequent abjuration. Rather than seeking evidence that Anastasia had been engaged in diabolic witchcraft, the inquisitors were primarily concerned with establishing whether she had practiced, in Herzig's words, "mere sorcery and superstition."[63]

Herzig's emphasis on diabolic witchcraft has led her to exaggerate the significance of these ideas in Anastasia's trial and to mischaracterize the cluster of Modenese trials as a witch hunt. It has also obscured deeper continuities in the history of the inquisition of magic. The remainder of the trials that occurred in Modena during this period were likewise primarily focused on charges of superstition and sorcery. To take one example, in January 1517, Lucrezia Pansini appeared before the inquisitor to present fresh allegations against Don Guglielmo, the priest who had featured in the 1499 trial of Stadera discussed in the previous chapter. Lucrezia explained that for some time she had been troubled by a demon in her right leg. She was subsequently advised to consult Don Guglielmo, who took her to his church, where he attempted to exorcise the demon by making her stand in a circle drawn on the ground before the altar while he hit her clothes "with a bundle of olive branches, reciting some words she [could not] remember." She claimed that the priest subsequently attempted to coerce her into sexual intercourse, stating that he had initially cursed her and would do so again if she did not submit to his advances.[64]

Summoned by the inquisitor, the priest cautiously confessed to a number of offenses, but his responses reveal a defensive strategy that was informed by a clear understanding of the boundaries of magic. He explained that, contrary to local belief, he did not keep a demon. He then admitted that he had given Lucrezia the impression that he could help her contact a demon but stressed that his activities were, in fact, an elaborate deception. To make her believe that she was communicating with a spirit, he had enlisted the help of a fellow priest who, hidden from view, had played the part of a demon and responded to her. Asked whether he had made

Lucrezia say prayers to the fake demon, he claimed that he had only encouraged her to say ordinary prayers before her conversation with the demon in order to set her at ease. In other words, he had not encouraged any superstitious activity.[65]

Although Don Guglielmo attempted to cast his actions as deceitful rather than superstitious, over the course of several interrogations he conceded that he had previously possessed and read various works of magic, including the *Clavicula salomonis*. He also admitted that he had been pressed by parishioners to perform various forms of healing and love magic on their behalf. For example, pestered by local women, he had produced magical slips designed to engender love. He claimed, however, that he merely wrote on them "the first words that would come to his mind" and never diabolical characters or the names of demons. Although Don Guglielmo again presented his actions as fraudulent, he nevertheless demonstrated an awareness of which practices—here the act of creating talismans by inscribing words or symbols on a given material—might be of concern to inquisitors. During his final interrogation, Don Guglielmo admitted to having once kept a familiar spirit in a crystal ball. Defending his actions, he claimed that he had commanded the demon using "exorcisms that are commonly used against demons." His argument may have been entirely ingenuous and certainly reflected the arguments of some magicians, but as we have seen, it was one that most inquisitors would have rejected. The priest was subsequently convicted of heresy on various grounds, including keeping a demon and producing slips of paper inscribed with markings that the inquisitor determined were designed to solicit demons. Don Guglielmo subsequently appealed to the Apostolic Penitentiary, which overturned the inquisitor's verdict and permitted him to resume his duties.[66]

The evidence from these trials suggests that in the late 1510s and early 1520s, inquisitors in Modena and indeed members of Modenese society were largely uninterested in pursuing the witch beliefs developed by Nider and Institoris and Sprenger. There is only limited evidence in these investigations to suggest that the inquisitor was seeking evidence that the accused were part of (or even knew about) a sect of devil-worshippers conspiring to perform *maleficia*. In the city of Modena at least, individual inquisitors active at this time prosecuted "mere" superstition and sorcery with unprecedented enthusiasm. Equally, there is no evidence in these trials to suggest that the inquisitors' actions were informed by the renewed Thomist conception of superstition that was also evident in their writings. Although at Lateran V some intellectuals such as Giustiniani and Querini had called for the condemnation of ostensibly natural arts and any works in which they were described, these ideas did not influence the conciliar decree. The evidence from these trials also shows that inquisitors were unaffected by these developments. During their investigations, the Modenese inquisitors did not consider whether the accused had performed an art of contested orthodoxy, nor did their investigations turn on the question of whether any particular practice was naturally possible. Instead, the cases were settled by finding evidence that the accused had performed ritual magic. In both of these senses, the trials conducted in Modena during the early sixteenth century were consistent with the principles for conducting the investigation of magic formulated by the papacy in the fourteenth century and subsequently set down by Eymerich in his inquisitorial handbook.

Conclusion

Lateran V closed on March 16, 1517. It had been a qualified success for the papacy. Leo X had seen off the threat of schism, secured the abrogation of the Pragmatic Sanction, quelled a potentially divisive dispute between the mendicant orders and the secular clergy, and passed a number of reform decrees. There nevertheless remained significant fault lines within the Church. Conciliarism, conceived both as a desire to use councils to effect reform and as the assertion of the supremacy of councils over the papacy, continued to be highly influential. Furthermore, the council had not managed to fully resolve several disputes related to conciliarism—namely, defining the relationship between the episcopate on one hand and the papacy and the mendicant orders on the other. Many within the Church also continued to call for further reform. While the fathers of Lateran V had set out a series of principles that would enable the episcopate to lead a regeneration of the Church, these ideas still needed to be implemented. There also remained rival visions of reform such as the mendicants' program of pastoral renewal. As we shall see, each of these issues resurfaced in subsequent years, following the outbreak of Europe's sixteenth-century religious crisis.

✷ 6 ✷

The Reformation

TRENT AND THE ESTABLISHMENT OF THE ROMAN INQUISITION, 1517–49

Introduction

At the beginning of the sixteenth century, there had been diverse calls for reform, conceived in multiple different ways. The fathers of the Fifth Lateran Council had laid out the basis for enacting a bishop-led renewal of pastoral care but had not discussed doctrine. Throughout this period, other members of the Church proposed alternative reforms, including changes to doctrine and religious practice. The mendicants too continued their ministry, promoting their conception of reform. The situation was complicated in 1517 by the crisis triggered by Martin Luther's protest. The Catholic Church's initial response was slow, but by the 1540s Pope Paul III had convoked a universal council and established the Roman Inquisition. They were not, however, created to serve a single common purpose. The Roman Inquisition was founded to contain heresy until the council could both define doctrine and outline a program on institutional reform. Both also became vehicles for expressing different reformist aspirations within the Church. The council was supported by those in the Church who favored the conciliar tradition of bishop-led reform, and several influential Tridentine fathers were amenable to doctrinal reform and dialogue with the Protestants. The Roman Inquisition expressed the views of a faction in the Church who upheld the papal supremacy, advocated that reform should be defined and led by the papacy, rejected doctrinal innovation, and opposed compromise with the Protestants. These views coincided with those of the mendicants who staffed the Roman Inquisition. The new institution offered them a means to promote their preferred response to Luther and to extend their program of reform, including the pursuit of superstition.

Luther's Protest

Toward the end of 1517, the curia received a copy of a work entitled *Disputatio pro declaratione virtutis indulgentiarum* (*Disputation of Martin Luther on the Power and Efficacy of Indulgences*). Written by an obscure Augustinian friar, Martin Luther, it contained a series of propositions for debate concerning the practice of selling indulgences. Luther's protest quickly broadened, and he began to challenge received doctrine and ecclesiastical structures.[1] From an early stage, the Holy Roman emperor Charles V (1500–1558) advocated the convocation of a council as a means to resolve the ensuing crisis. He reasoned that since abuses had triggered Luther's initial protest, resolving them might bring it to an end. Political circumstances rendered a council unlikely, however. The French king, Francis I, opposed the idea for strategic reasons. Fearing that his realm could be surrounded by Charles's sprawling territorial possessions, he calculated that the crisis would keep the emperor occupied. The incumbent pope, Clement VII (1522–34), in any case wary of councils, feared Charles's growing power in the peninsula. In 1526, he entered into a military alliance against Charles, the League of Cognac, with nations including France. The tense situation deteriorated in 1527, when the imperial army defeated the French forces in Italy. The imperial commander was unable to pay his troops, and they subsequently mutinied and sacked the city of Rome.[2]

Despite the upheavals of the 1520s, a number of individuals in Italy continued to advocate diverse types of reform. After the sack of Rome, the Oratory of Divine Love temporarily disbanded. Several of its members devoted themselves to earlier reformist ambitions. The bishops Jacopo Sadoleto (1477–1547) and Gian Matteo Giberti (1495–1543), for example, returned to their dioceses, where they began to act as model bishops. Residing in their sees, they disciplined clergy, reformed religious houses, and devoted themselves to the renewal of pastoral care. The oratory subsequently regrouped in Venice, where it drew together reformers of contrasting opinions. Carafa, for instance, continued to advocate reform and actively sought the spiritual renewal of the Church. He nevertheless explicitly rejected Luther's ideas and opposed the idea of dialogue with the Protestants. Reflecting the diversity of opinion within this community, other members of the oratory, such as Marcantonio Flaminio (1497/8–1550), openly sympathized with Lutheran views.[3] By as early as 1523, another member of the oratory, Gasparo Contarini, had read some of Luther's works. Their influence was evident in a letter that he wrote to Paolo Giustiniani, in which he suggested that he now rejected the role of works in salvation. Although Contarini remained unwavering in his obedience to the papacy, he and other like-minded individuals continued to believe that it might be possible to draw on Luther's thought to inspire reform within the Catholic Church.[4]

Although several influential figures in Christian society continued to believe that it might be possible to accommodate Lutheran ideas within Catholicism, the divisions between the two faiths were becoming increasingly clear. By the late 1520s, German Protestant princes had formed a defensive organization, the Schmalkaldic league, which afforded protection to Luther and his followers, enabling them to form a new Protestant state-church.[5] Meanwhile, opinion in Spain began to harden against humanist reformers and especially Erasmus. When Charles left Spain for Italy in 1529, the opponents of humanist reform—Dominican and Franciscan friars prominent among them—gained control of the Spanish Inquisition and began to extinguish Erasmian traditions there.[6] The threat of an inquisitorial investigation also forced one Spanish humanist, Juan de Valdés (1490–1541), a nobleman and proponent of doctrinal reform, to leave his homeland. He traveled to Italy, where he continued to discuss his ideas—an eclectic blend of Lutheranism, Erasmianism, and mysticism—relatively openly.[7] Despite the Dominicans' efforts, Erasmus's influence lived on in the Society of Jesus. It was founded in 1542 by a Spanish nobleman, Ignatius Loyola (1491–1556), who had studied at the University of Alcalá.[8]

In the early 1530s, Juan de Valdés settled in Naples, where soon he gathered around him a network that included Bernardino Ochino, Peter Martire Vermigli, Vittoria Colonna, and Giulia Gonzaga.[9] His arrival broadly coincided with and encouraged the further development of the ideas of a network of Italian evangelicals, who are often referred to as the *spirituali*.[10] In addition to Valdés's Neapolitan followers, many of the *spirituali* were drawn from a network of reformers resident in the Veneto. They included members of the Oratory of Divine Love such as Contarini and Flaminio, theologians from the Benedictine Congregation of Santa Justina and Santa Maggiore, and the English cardinal Reginald Pole. Throughout the 1530s, members of this network continued to engage with the theological writings of northern Reformers and embraced a number of their doctrines, most notably justification *sola fide*. Like the emperor, they also supported calls for dialogue with the Lutherans. They believed that the Catholic Church should draw on the Reformers' ideas to effect a doctrinal reform, which would be approved by a general council. If this were achieved, they maintained, it would be possible to heal the schism within the Church.[11]

During these years, Clement recognized that the papacy needed to take more determined action to control the growing crisis. He continued to resist calls to convoke a council, however. Although his predecessor, Leo X, had successfully secured the abrogation of the Pragmatic Sanction and passed a decree at Lateran V restraining the strict forms of conciliar theory that asserted the council's authority over the papacy, Clement continued to fear these ideas. Furthermore, following the sack of Rome, he was ill-disposed toward Charles V. He was also coming under increasing pressure from members of the Church who rejected the idea of compromise with the Lutherans. In 1532, Carafa wrote to Clement, reminding him that "a heretic must be treated as such" and urging him not to enter into dialogue with those who held false beliefs. His remarks set the tone for the debates that would follow about how best to respond to the Lutheran challenge. While some continued to argue that it was both necessary and beneficial to engage with the Lutherans and perhaps reach some form of compromise that could heal the schism, Carafa sought only to eradicate their views from the Church.[12]

The Inquisition of Heresy in Italy, 1520–41

Despite the spread of Lutheran ideas in the Italian peninsula during the 1520s, it appears that inquisitors of heresy did not expend any great effort to prosecute Protestants or philo-Protestants. This lax approach to Lutheranism formed part of a broader picture of inquisitorial inactivity. In his study of Dominican inquisitors in northern Italy, Michael Tavuzzi has shown that when investigating heresies such as Waldensianism or cases of alleged Judaizing, inquisitors tended to go about their business in a somewhat desultory manner. Del Col reaffirmed this impression, noting that in Venice, for example, there were very few trials for heresy between the mid-1520s and mid-1530s. The inquisitors may also have been taking their lead from the papacy, which remained relatively uninterested in Lutheranism through the 1520s. By the early 1530s, however, Clement VII began to recognize the need to find more efficient ways to investigate and prosecute those who held recognized heresies. He began to use papal nuncios to investigate cases of alleged heresy in Venice and, to a lesser extent, Savoy and Naples. In other cities, local bishops or their vicars led the fight against heresy. In 1532 the papacy attempted to coordinate efforts to repress Lutheranism in Italy through the appointment of an inquisitor general, but little is known about how this office functioned.[13]

As Tavuzzi has observed, the only major exception to this pattern of inquisitorial activity in the 1520s was prosecutions for witchcraft and superstition.[14] Indeed, the 1520s witnessed the largest series of trials yet seen in Italy. Between 1518 and 1521, the bishop of Brescia and the local inquisitor executed between sixty-two and eighty men on charges of performing diabolical witchcraft and investigated at least as many more. In 1523, the inquisitor of Como, Modesto Scrofa da Vicenza, oversaw the trial of around thirty people in Sondrio, which led to the execution of at least seven. Finally, between 1522 and 1523 the inquisitor of Parma and Reggio led an investigation of sixty people in Mirandola that resulted in ten executions. While horrifying, these large-scale trials were the exception rather than the rule. During the 1520s, there were a further six trials across all tribunals. They were relatively small, with the number of accused varying between one and seven people. There is currently no evidence for any large-scale hunts during the 1530s. If anything, there appears to have been a decline in prosecutions for magic. According to Del Col, over the course of this decade, inquisitors conducted two trials, during which they tried just nineteen people. There was one further trial conducted by a secular tribunal in Siena, which tried a further four individuals.[15] The available evidence therefore indicates that the phenomenon of large-scale trials had ended by the 1530s.

The conclusion that the last mass trials for magic and witchcraft in Italy occurred in the 1520s calls into question the historiographical commonplace that the witch hunts were brought to a close by the foundation of the Roman Inquisition.[16] More specifically, it problematizes John Tedeschi's thesis that the foundation of the Roman Inquisition in 1542 ended the witch trials because its centralized structures and appellate structure prevented local officers from abandoning due process.[17] Tedeschi's thesis, in any case, rested on the implied premise that the earlier witch hunts were able to occur because, before 1542, inquisitors were not bound by these institutional safeguards. Recent research has questioned this assumption. Notably, Tavuzzi concluded that although the inquisitors who conducted the trials in Valcamonica in 1518–21 and Sondrio in 1523 may indeed have abandoned due process, such actions do "not seem to be the norm."[18] Furthermore, there is no reason to suggest that before the establishment of the Roman Inquisition accusations of diabolism necessarily led to mass investigations. There were, in fact, several documented instances of inquisitors prosecuting small numbers of individuals for diabolism during the 1520s and 1530s that did not escalate into mass hunts.[19]

Let us examine one case in more detail. In 1531, three witnesses made a collective deposition to the inquisitor of Modena—who, like the inquisitor of Como, was drawn from the Congregation of Lombardy—alleging that a certain Brighento Brighenti belonged to the sect of witches and that he was also a sorcerer who used his power not only to cure but also to bewitch humans and animals. Investigating these claims, the inquisitor enquired about his healing abilities and the nature of his treatments for both humans and animals. He replied that he cured them by natural means acquired from an apothecary. The inquisitor also asked whether he believed that witches or sorcerers could use magic to work either beneficent or maleficiant ends. This was a potentially dangerous question, for the authors of the *Malleus maleficarum* had argued that to deny the truth of magic was to entertain a heresy. Brighenti nevertheless stated that he did not believe that witches and sorcerers possessed these powers. His denial prompted the inquisitor to ask whether the friars were just to burn witches, to which he replied that he did not know whether it was fair. In answer to further questions, he denied having ever attended the Sabbath or describing himself as a witch. He also rejected any claim that he used any magical healing techniques, but he admitted reciting prayers over those afflicted by hernias. Brighenti was subsequently found to be "lightly suspect" of the "heresy of the witches." He was forced to abjure his beliefs and to refrain from using superstitious healing techniques in the future.[20]

Brighenti's case illustrates three important points. First, it shows not only that inquisitors continued to harbor concerns about the existence of a diabolical sect and to question individuals about it in the 1530s (that is, after mass hunts ended) but also that this idea had penetrated the imaginations of the laity in regions such as Modena that lay outside the epicenter of witch-hunting in the valleys of northern Italy. Second, it indicates that the presence of these beliefs among inquisitors and the population at large did not necessarily lead to unduly harsh punishment; Brighenti abjured his offenses but was not subject to corporal, let alone capital, punishment. A further case suggests that this relative leniency was not exceptional. In 1539, a woman named Orsolina admitted under torture offenses including attending diabolic Sabbaths and—in an echo of the earlier Italian mythology of the *strega*—sucking the blood of infants. She was convicted and sentenced to house arrest for the remainder of her life.[21] Finally, Brighenti's case demonstrates that inquisitors' investigations of allegations of diabolic witchcraft did not necessarily trigger further witch hunts. It seems that inquisitors did not require the restraining hand of central authority to prevent them from embarking on a program of mass persecution. This conclusion in turn suggests that we should not seek to explain why members of the Roman Inquisition did not preside over large-scale witch hunts. Instead, it may be more productive to identify why some inquisitors were able to abandon due process and inflict such horrors between circa 1460 and circa 1490 and once again in the 1510s and 1520s. It also suggests that to explain the decline of mass trials, we may need to examine local factors—for instance, the protests of local clerical and secular leaders.[22]

This discussion of Brighenti's trial also indicates important continuities in the inquisition of magic. Although the Modenese inquisitors did question Brighenti about diabolism, they were also concerned—perhaps more deeply—by his superstitious healing practices. These documents also indicate one further important area of continuity in inquisitorial practice. While the Modenese trial documents record frequent examples of individuals being accused of necromancy or love magic, there were still no examples of individuals being prosecuted for practicing ostensibly natural arts such as chiromancy. As we have seen, there had been a transformation in thinking about how to define superstition during the fifteenth century. These ideas had been adopted by some influential figures within the Church in the early sixteenth century. The Camaldolese hermits Querini and Giustiniani, for instance, had warned of the dangers posed by these arts. Members of the learned elite nevertheless continued to produce books describing them. To take one example, in 1525 Patricio Tricasso produced an Italian translation of the latter three books of della Rocca Cocles's *Anastasis,* which were concerned with chiromancy, and in 1538 he published *Epitome chyromantico,* a vernacular work on chiromancy. It may be that inquisitors simply did not encounter such works in the course of their duties. It is, however, equally possible that these inquisitors did not believe that these arts warranted their attention. The documents generated by inquisitorial trials conducted in Modena during this period certainly offer no evidence to suggest that the Thomist criteria for defining the boundaries of operative arts articulated by Nider, Institoris, and Sprenger had displaced Eymerich's emphasis on finding tangible evidence of invocation in inquisitorial practice.[23]

Dealing with Luther: Inquisition or Council?

Clement VII died on September 25, 1534. He was succeeded by Cardinal Alessandro Farnese (1520–89), who assumed the name Paul III. Upon his elevation, Paul announced three interrelated aims for his pontificate: establishing peace among Christian princes, convoking a council, and leading a crusade against the Turks. By promising a council, Paul had set himself the daunting task of maintaining the unity of a doctrinally divided Church, responding to calls for institutional reform, and brokering agreements between the princes of Europe. His task was further complicated by the need to guard against any potential challenges to papal authority. An emboldened episcopate might seek further reforms that strengthened their power relative to the papacy, while both bishops and secular rulers might seek to use strict forms of conciliar theory to advance their interests.[24] Despite the risks, Paul decided to proceed, but his first attempt to convoke a council in Mantua in 1536 ended in failure. He nevertheless created a commission to consider ecclesiastical reform, to which he appointed the evangelical cardinals Reginald Pole and Gasparo Contarini and traditionalist cardinals Gianpietro Carafa and Girolamo Aleandro.[25] Their report, *Consilium de emendanda ecclesia* (1537), did not touch on doctrine but proposed changes—such as improving the quality of candidates for the priesthood and raising their levels of education, reforming the administration of benefices, and encouraging episcopal residence—that carried forward the program of pastoral reform outlined at the Fifth Lateran Council. They also called for reform of the religious orders; specifically, they suggested that conventual orders should be abolished and that friars engaged in the cure of souls be made subject to the local ordinary.[26]

Meanwhile, tensions between the evangelical and traditionalist wings of the Church continued to intensify. They were brought to a head at the Colloquy of Regensburg, a new initiative designed to allow Protestants and Catholics to bridge their doctrinal differences. At this meeting the Catholic representative, Gasparo Contarini, proposed a twofold theory of grace, a compromise that he hoped would resolve the disputed issue of justification. His suggestion was rejected by Protestants and Catholics alike, and the meeting ended in failure.[27] Contarini returned to Italy in autumn 1541, whereupon Paul III appointed him to the Legation of Bologna. This was a concerning development for Carafa, for this role invested Contarini, a man whom he suspected of holding heterodox beliefs, with responsibility for coordinating the campaign against heresy in northern Italy. To make matters worse from Carafa's perspective, Contarini was supported in this work by a number of his *spirituali* associates, including Giovanni Morone, the bishop of Modena; Gian Matteo Giberti, the bishop of Verona; and Cardinal Pole, the governor of the largest of the Papal States. Not only did they seek to tackle heresy by persuasion rather than force, but they did so by engaging the services of preachers such as Bernardino Ochino and Peter Martyr Vermigli who were associated with Juan de Valdés. After Valdés's death in 1541, Pole's residence in Viterbo became the center of the *spirituali*'s activities. Carnesecchi and Flaminio lived in Pole's household, while Vittoria Colonna stayed in the nearby convent of Santa Caterina. At this time, the circle in Viterbo produced one of the defining texts of the so-called Italian Reformation, the *Beneficio di Christo*, which blended elements of Italian evangelism with Valdés's thought and the ideas of John Calvin.[28]

The failure of the Colloquy of Regensburg convinced Paul III that reconciliation with the Lutherans was unlikely, if not impossible. These revelations prompted Paul to change his approach to the religious crisis. Although he remained committed to convoking a general council, he now conceived it as an opportunity for Catholics to define the orthodox parameters of their faith rather than to negotiate with the Lutherans. On June 29, 1542, Paul issued a bull convoking a council in the city of Trent, a compromise location that he hoped would be acceptable to both the emperor and the king of France. The outbreak of war between the two princes forced Paul to suspend the council in September 1543.[29] Meanwhile, reports of increased incidences of heresy in Venice and Modena persuaded the pope that existing efforts to combat Lutheranism in Italy were insufficient.[30] Seeking alternative means to contain the spread of heresy, Paul appointed Cardinals Carafa and Aleandro to reform the practice of inquisition. Following their proposals, Paul formally instituted the Roman Inquisition with the passage of the bull *Licet ab initio* issued on July 21, 1542. Its text suggested that the pope intended the new institution to be a temporary expedient, one that could enable the efficient prosecution of heresy until a council had met and successfully concluded its business.[31]

Taking the Spanish Inquisition as an organizational model, the new institution was administered by the Holy Office, a centralized bureaucracy headed by six cardinals that controlled a network of local tribunals staffed primarily by members of the mendicant orders. The new institution had the potential to recalibrate the balance of power between the papacy and the episcopate on the one hand and between traditionalists and evangelicals on the other. Although the power of absolution and reconciliation remained the preserve of the papacy, Paul III invested inquisitors with the authority to pursue heresy throughout Christian society, irrespective of rank or position.[32] The pope enhanced these powers in July 1543 by issuing an edict that redefined the system of censorship established at Lateran V. It allowed inquisitors to bypass bishops in order to directly inspect libraries, printing works, private houses, and monasteries and to seize and burn any heretical works that they found. In principle at least, these new powers allowed the Holy Office to centrally coordinate the investigation and censorship of potentially heterodox materials.[33]

In theory, the Roman Inquisition's power was significant, but in practice it was limited. This situation was partly caused by inconsistencies in the new institution's local organization. The papacy could establish local tribunals only with the permission of secular rulers, creating considerable regional variation. For example, the Habsburgs refused to allow the Roman Inquisition to be established in the kingdom of Naples, while the papacy denied the Habsburgs permission to introduce the Spanish Inquisition there. The result was an awkward compromise. As Christopher

Black has noted, "The Kingdom did have certain inquisitorial mechanisms linking the Congregation of the Holy Office to archbishops and bishops, who had commissioners and other officials with inquisitorial authority, rather like the old medieval systems." In those states where the Roman Inquisition was introduced, local inquisitors performed their task within complex networks of competing jurisdictions. In Venice, for example, the tribunal's powers were checked by local religious and secular authorities. Even in the Papal States, inquisitors found that they struggled to assert their authority against city governments, local bishops, and the vicars that they appointed to prosecute heresy.[34]

The Inquisition's pursuit of Protestantism was further complicated by the fact that since orthodox doctrine remained undefined, it was difficult to identify which precise ideas should be considered heretical.[35] The complex situation was exemplified by disputes about the *Beneficio di Christo*. In 1543, the *spirituali* greeted the publication of the second edition of this work with enthusiasm. Morone, for example, encouraged its circulation within his diocese. Members of the Inquisition urged its suppression, however. A Dominican theologian, Ambrogio Catharino, subsequently produced a rebuttal of the work's theological propositions, including its arguments concerning the doctrine of justification. As Dermot Fenlon has argued, the publication of Catharino's text "illustrates the growing awareness of the need to define and clarify the guidelines of belief among Italian Catholics and, through a General Council, among Catholics at large."[36]

The First Period of Trent

For some years, convening a universal council had appeared an unattainable goal. The situation changed in 1544, when Charles V secured a stunning military victory over Francis I. As part of the Peace of Crépy, Francis agreed to the French Church's participation in the council. It finally opened on December 13, 1545. To retain control over the meeting, Paul III invested his legates, Cardinals Pole, Marcello Cervini, and Giovanni Maria Del Monte, with the power to preside over the council and set its agenda. Before the first session, the assembled fathers held a General Congregation to discuss the procedures that the council would follow. It rapidly became apparent that the emperor and the pope held conflicting ambitions. Pursuing what Hubert Jedin has dubbed his "Great Plan," Charles V wanted to prioritize discussion of reform. He hoped that implementing reforms might allay some of the Lutherans' concerns while delaying discussion of doctrine until they could be either persuaded or compelled to attend the council. For his part, Paul III wanted to avoid discussion of reform lest it necessitate changes within the curia. Furthermore, confirmed in his opinion that Germany was lost to the faith, he saw the council primarily as an opportunity to define Catholic doctrine. To break the impasse, the council fathers reached an awkward compromise by agreeing to discuss doctrine and reform simultaneously.[37]

Proceeding in this manner, the council fathers began to treat substantive issues. During Session Four, held on April 8, 1546, the assembled fathers accepted the council's first two decrees: "On Sacred Books and Apostolic Traditions" and a reform decree, "On the Vulgate Edition and on the Proper Interpretation of Scripture." The first established the canonical works of scripture and the role of Catholic traditions in its interpretation. The second affirmed that the Vulgate was to be regarded as the "authentic" version of the text of scripture. It did not, however, comment on an issue that would later play a significant role in post-Tridentine Italy: the legitimacy of vernacular translations of the Bible. Having successfully passed these decrees, the fathers began to prepare for the following sessions. Seeking to address the doctrinal issues that lay at the heart of the religious crisis, they proposed to discuss the doctrine of original sin and justification. The council fathers also began to consider ideas for inclusion in two reform decrees. The first recommended that, where feasible, scriptural lectureships should be established in dioceses and religious houses, including in those of the mendicants. The second decree was concerned with preaching.[38]

Discussion of the reform decrees offered a foretaste of some of the most contentious battles of Trent, which were concerned not with doctrine but with papal authority. Since the proposed decree on preaching encouraged bishops' presence in their dioceses, it indirectly raised the disputed issue of episcopal residence. For reformers, enforcing residence offered a means to rebuild pastoral care at the diocesan level, but the papacy feared that it could curtail their powers of patronage. Furthermore, drafts of the decree on preaching declared that while the friars could preach in their own churches with a license from their superior, in order to preach in parish churches they should first obtain the permission of both their superior and the bishop. This decree therefore continued the efforts made at Lateran V to limit the papacy's power to grant the mendicants privileges to conduct the cure of souls without the consent of the local ordinary. While discussion of these issues was more peaceful than it had been thirty years previously, during the General Congregation the bishop of Fiesole, Braccio Martelli (1501–60), angrily denounced the friars' privileges. For his part, the legate Del Monte considered these complaints an attack on papal authority. Despite the tensions, Session 5 was held on June 17, 1546, during which the council fathers successfully passed each decree.[39]

The Tridentine fathers now began to prepare for the next session, which would discuss a further reform decree and the contentious issue of justification. The emperor continued to hope that it might still be possible to reach a compromise with the Protestants on this fundamental question. Furthermore, several influential members of the Catholic hierarchy remained sympathetic to the doctrinal principle of justification *sola fide*. Prior to the General Congregation, Pole urged the fathers not to dismiss the Lutherans' views out of hand. Discussion of these delicate matters was passionate, at times heated, and on one notable occasion even descended into physical violence. A further division within the council opened when, upon hearing that the Lutherans posed a military threat to the city of Trent, the papal legates proposed to move the council further south. The fathers subsequently coalesced into two factions: an "imperialist" party of bishops drawn mainly from Spain and the kingdom of Naples who resisted the translation of the council, and a "curial" party consisting of bishops from the curia and the Papal States who supported the legates' position.[40]

These lingering tensions were exposed once more when the fathers began to discuss the proposed reform decree, which was concerned with episcopal residence. In their desire to finalize the text of the decrees on justification in time for a session on January 13, the legates had left little time in the General Congregation for consideration of this contentious issue. Despite the limited time available, reform-minded fathers, drawn primarily from the imperial party, wanted to discuss the idea—first proposed by Cardinal Cajetan in 1517—that episcopal residence should be considered an obligation *de jure divino*—that is, according to divine law. These discussions held important ecclesiological implications, for they raised the question of whether bishops received their authority directly from God, via their consecration, or from the pope. This question had underpinned the debates of the conciliarist period and was connected to disputes about the papacy's authority to grant privileges to the mendicant orders. The imperialists were supported in their efforts to discuss the principle of residence *de jure divino* by a limited number of Italian bishops, several of whom were associated with the *spirituali*. It was, however, rejected by members of the curialist party, who recognized the threat that it posed to the papal supremacy. The legates subsequently proposed limited reforms, including fining bishops if they were absent from their diocese without legitimate cause, but they avoided not only the thorny question of papal dispensations that allowed bishops to hold multiple benefices but also discussion of the principle of *jus divinum*. The "imperialist" party were scathing about these proposals, arguing that they did little to encourage the bishop's pastoral role.[41]

As planned, Session 6 was held on January 13, 1547. With the approval of the emperor, the fathers passed the final decree on justification, which reaffirmed the existence of free will but clearly stated that, contrary to Lutheran belief, original sin was remitted by baptism. With regard to justification, the council opted for a position that emphasized the priority of grace in salvation but that nonetheless guaranteed a limited role for human endeavor. The adoption of these ideas, coupled with the rejection of specific Lutheran beliefs, made the prospect of reconciliation with the Protestants far less likely. It was also a significant defeat for members of the *spirituali*, for it anathematized several of their key beliefs. Recognition of this fact may have been why Pole, ostensibly on the grounds of ill health, chose to withdraw from the council before the final vote on the decree concerning justification. During this same session, the fathers also narrowly passed the reform decree. The reformers had won some concessions, notably securing the imposition of stronger penalties on absentee bishops, but the legates had successfully avoided discussing the idea that residence was required *de jure divino*. It nevertheless remained clear to the legates that they would need to return once more to the issue of episcopal residence.[42]

On February 3, 1547, the fathers opened a General Congregation to discuss proposals for a further reform decree dealing with residence and doctrinal decrees relating to the sacraments. The reform decree was presented to the council during Session 7, held on March 3. Its fifteen chapters were designed to ensure the appointment of suitable candidates who were committed to fulfilling a bishop's duty to attend to the cure of souls. It stated that bishops should hold only one bishopric and required them to conduct annual visitation of each church in their diocese to ensure both the quality of pastoral care and the physical structure of the churches. Although reformers wanted more radical change, as John W. O'Malley has noted, "the decree contained the nucleus of the Tridentine reform of pastoral offices." This session also passed a doctrinal decree, which reaffirmed that there were seven sacraments, each necessary for salvation. The decree also included canons on baptism and confirmation. Once again, while this decree helped define and clarify orthodox Catholic belief, it also served to further entrench divisions with the Protestants.[43]

The fathers laid plans to hold a further session on April 21, which would discuss the Eucharist. It seemed that the council was moving to a conclusion and could perhaps complete its work by the following May or June. These hopes were dashed on March 8, when Del Monte informed the fathers of a suspected outbreak of plague in Trent and proposed to move the council. The majority of the fathers voted to translate the meeting to Bologna, but the decision was opposed by Charles V and the imperialist party at Trent, who believed that the emperor was on the brink of realizing his Great Plan. They feared that although Charles could secure his military objectives, his wider strategy might unravel if the meeting were translated to the Papal States because it would give the Lutherans a reason to refuse to attend. On April 24, the emperor finally secured a decisive victory over the Schmalkaldic league and demanded that the council return to Trent. While Charles's victory pleased Paul III, it also reaffirmed his fear of Habsburg power, and he refused to accede to his requests. In May 1547, the fathers resumed their deliberations in Bologna, beginning with discussion of matters relating to the Eucharist such as the real presence of Christ, the doctrine of transubstantiation, and taking mass in both forms. The fathers also began to discuss the sacrament of penance, which raised related contested issues such as purgatory and indulgences, extreme unction, holy orders, and marriage. The council nevertheless delayed holding a further session, preferring to wait until the emperor and the king of France recognized the meeting. After two years of waiting, Paul finally chose to dissolve the council rather than formally adjourn or conclude it.[44]

The Legacy of Trent's First Session

The suspended council left an ambiguous legacy. It had established a limited set of reform decrees, albeit ones that failed to satisfy many in the Church. Its doctrinal decrees had done a great deal to clarify orthodox belief, especially with regard to justification, although a number of questions regarding the sacraments had not been resolved. Charles still hoped that it would be possible to secure a lasting compromise with the Protestants, but the council fathers' decision to champion traditional doctrine made this an unlikely prospect. The passage of this decree also required the *spirituali* and other evangelicals to reassess their views on matters such as justification *sola fide*. *Spirituali* such as Cardinals Pole and Morone nevertheless continued to occupy influential roles in the ecclesiastical hierarchy. As Camilla Russell has argued, while they no longer became involved in theological disputes, they continued to concern themselves with debates about Church reform, especially the disputes over the principle of requiring episcopal residence *de jure divino*. This idea, she noted, "appealed to the *spirituali*, who identified Scripture, rather than tradition, as the animating force of the church, as well as its primary authority."[45] More broadly, evangelical ideas remained influential. The council had not passed judgment on key tenets of evangelical thought such as widening access to scripture, and the writings of thinkers such as Desiderius Erasmus continued to inform the ideas and practices of groups within the Church.

The council's decrees also provided clearer guidance to the Roman Inquisition, which was now controlled by four doctrinal conservatives, Cardinals Carafa, Toledo, Sfondrati, and Cervini. Under Paul III's moderating influence, the first years of the institution's activities had been characterized by relative restraint. Although two important evangelicals, Ochino and Vermigli, chose to flee Italy in the early 1540s, inquisitorial investigations for Lutheranism were rare. By the end of the decade, the Roman Inquisition began to take a more proactive approach to the prosecution of heresy. This was partly because Trent's doctrinal decrees had provided a clear definition of the dividing line between orthodox and heterodox beliefs. The Roman Inquisition's activities were also prompted by concerns about evangelical reformers within the ecclesiastical hierarchy. The Tridentine decrees had not yet been ratified by the pope and could, in principle, be amended if the council were resumed. High-ranking members of the Roman Inquisition sought to safeguard established doctrine by attacking those who might later seek its revision. In the late 1540s, the Roman Inquisition opened investigations into several leading clerics sympathetic toward evangelical ideas including Pole, Morone, and the nobleman Pietro Carnesecchi. To Carafa's irritation, these investigations were curtailed, probably by Paul III.[46]

While the first session of the council made significant steps toward defining Catholic doctrine and formulating a program of episcopal reform, it was also notable for what the fathers did not discuss. Unlike the reform decrees of Lateran V, those of the first period of Trent did not mention superstitious magic. How, if at all, did this omission affect inquisitorial activity? We do not currently possess statistics for rates of inquisition specific to the first ten years of the Roman Inquisition's activities as a whole, but some data is available for the activities of the tribunals in Modena and Venice. Reflecting the moderate approach to the inquisition of heresy advocated by Paul III, investigations in both cities were limited. Between 1542 and 1550, there may have been as few as thirty-one inquisitorial trials in Venice, of which only one was concerned with magic.[47] Overall, Modena's tribunal was less active than it had been in earlier periods. In 1517–21, inquisitors had prosecuted thirty trials for magic alone, whereas they conducted only twenty-eight trials for all types of heresy between 1540 and 1549. Even at this time of heightened concern about the threat of Protestantism, it seems that inquisitorial activity depended on the zeal and effectiveness of the incumbent inquisitor as much as anything else. It is, however, noteworthy that of the twenty-eight trials conducted in this period, four were related to magic. The impetus for these prosecutions came neither from Carafa and his associates nor from the Council of Trent but from the Dominican inquisitors' continued engagement with the mendicants' program of reform. The Lutheran crisis may have provided a new focus for inquisitorial activity, but the pursuit of this heresy formed only a part of the mendicants' ongoing campaign to reform Christian society.[48]

Conclusion

The Lutheran crisis exposed a number of fault lines within the Church. It revealed not only disputes about the correct definition of doctrine but also ongoing debates about the correct ordering of the relationship between the episcopacy and the papacy, mendicant privileges, and questions about which institutions in the Church carried responsibility for defining and implementing reform. These debates came to the fore during the conciliar crisis, but they dated back at least as far as the thirteenth century. While for some in the Church, the Council of Trent offered a means to resolve these issues, others (notably Carafa and the mendicants) were skeptical of its value and sought alternative means to combat heresy and effect reform within the Church. For their part, the mendicants continued their ongoing program of reform, which involved the suppression of heterodox ideas and practices and the promotion of approved models of behavior. As we shall see in the following chapter, the Roman Inquisition provided them with a vehicle for these aspirations.

✷ 7 ✷

Between Trent and the Roman Inquisition, 1549–64

Introduction

At the end of the 1540s, there were two distinct reforming parties in the Catholic Church. With the support of the emperor, the first group favored the resumption of the council and remained open to the prospect of negotiation with the Protestants. The second party, led by Carafa, opposed conciliar reform and sought a more aggressive approach to Protestantism. Moreover, Carafa sought to use the Roman Inquisition as a means to promote his own agenda. Despite Carafa's opposition, the council met for a second session, but it again collapsed before it could conclude. The subsequent election of Carafa as Paul IV ended hope of a council and signaled the beginning of a new era of papal reform. The new pope sought to impose reforms on the episcopate while using the Roman Inquisition to attack those within the Church whom he opposed. During this period, the power of the Inquisition expanded greatly, supported by the development of new forms of censorship such as the Index of Forbidden Books. The Dominicans took this opportunity to advance their own authority. As we shall see, they embedded their reforming agenda into the first papal Index of 1559, including a new approach to the censorship of learned magic rooted in the thought of Thomas Aquinas. After the death of Carafa, the proponents of conciliar reform managed to secure the reconvocation of the Council of Trent. At the council, the fathers reasserted the authority of the episcopate and implemented measures to restrict the Roman Inquisition's power, such as limiting its influence over censorship. As part of this process, a Tridentine commission revised the 1559 Index and drafted new rules for the practice of censorship. Although the new index relaxed a number of the 1559 list's prohibitions, it retained the earlier list's innovative approach to the censorship of magic.

Julius III and the Second Period of Trent

Paul III died on November 10, 1549. The conclave held to elect his successor was shaped by political and doctrinal disputes. Charles V favored a candidate willing not only to reconvene the council but also to enter into serious discussion with the Lutherans. The new French king, Henry II (1547–59), opposed the council's resumption. He had continued Francis I's conflict with Charles and, like his father, feared that if a council could successfully heal the schism, the emperor would be able to devote more time to war with France. During the conclave, Cardinal Reginald Pole—known for his evangelical sympathies—emerged as Charles's preferred candidate and seemed likely to be elected. The French cardinals, seeking to prevent Charles from gaining a potential advantage, vehemently opposed his candidature. They were supported by, among others, Cardinal Carafa, who harbored suspicions about Pole's orthodoxy. In a dramatic intervention, Carafa denounced Pole as a heretic before a crucial vote in the conclave. Deadlock ensued. During this period, the former Tridentine legate Cardinal Del Monte emerged as a compromise candidate. He was elected as pope on February 7, 1550, taking the name Julius III.[1]

In the face of opposition from France, Julius reopened the Council of Trent in May 1551. To keep proceedings on track and deflect any attempts to use the council to impose reform on the curia, he appointed a single legate, Cardinal Marcello Crecenzio (1500–1552). On September 1, Henry II's representatives communicated his intention to boycott the council. His protests notwithstanding, the fathers pressed on. During the next session, held on October 11, they considered decrees relating to the sacraments, including the Eucharist. The fathers affirmed several propositions, such as the doctrine of transubstantiation, which served once more to deepen the divisions between Catholicism and Lutheranism. To keep the potential for dialogue alive, the fathers passed a further decree postponing judgment on four further articles primarily concerned with receiving the Eucharist in both kinds—a matter of significant concern in Germany—until the Lutherans arrived. As a further inducement for the Lutherans to attend the council, the fathers passed a decree granting them safe conduct. The fathers also passed a decree on reform. However, it avoided many of the issues of central concern to Spanish and German bishops, notably episcopal residence, which had been only partially addressed hitherto.[2]

In late October, the council fathers began to prepare for Session 14. Scheduled to be held in November, it was due to discuss the sacraments of penance and extreme unction and the contested issue of reform. In preparation for the meeting, the Spanish party presented a memorandum outlining a series of potential reforms. Crecenzio, the papal legate, rejected this document, stressing that only he had the power to table material for discussion. To the irritation of the Spanish party, when Crecenzio presented his draft reform decree to the General Congregation for discussion, it offered only limited suggestions and made no reference to their proposals. Session 14 was finally held on November 25, during which the fathers passed a decree on penance.[3] Latent tensions between the imperial and papal parties were once more exposed during preparations for the next session. It was scheduled to be held on January 25, and the fathers expected to discuss decrees concerning the Eucharist and sacrament of holy orders. The latter decree was designed to rebut the Protestant assertion that bishops were not superior to other priests. To achieve this aim, canon 7 of the draft decree stated that bishops were appointed *de jure divino*—that is, by divine law. The insertion of this idea was contentious because it implied that, like the popes, bishops derived their authority directly from God. Several reforming bishops supported such an ecclesiology, hoping that it would enhance the authority of their office, but other clerics feared that it could undermine papal supremacy. While the legate permitted the inclusion of this idea in the draft decree, several Spanish bishops began to suspect that he was also attempting to insert a form of words that asserted the precedence of the papacy over the council.[4]

After the arrival of a delegation of Lutherans and the ensuing negotiations over the terms of their potential participation in the council, the fathers agreed at Session 15 to postpone discussion of the matters at hand. The situation was further complicated by political events that had been in train since the beginning of the year. On January 15, the Lutherans had concluded an alliance with Henry II. Soon, they were once more in open revolt in Germany. Seeing his Great Plan begin to unravel and frustrated that proposals for reform were being ignored, Charles V came to believe that the council's suspension was highly likely. The Lutherans subsequently secured a military victory over Charles's forces, which resulted in the capture of Augsburg. From this position, Lutheran armies could threaten Trent. After these events, the Tridentine fathers voted to suspend the council for two years. There followed a series of political and religious realignments in Europe. Charles V concluded the Peace of Augsburg with the Lutherans in September 1555. Its most significant provision was enshrining in imperial law the principle of cuius regio, eius religio—that the religion practiced in each of the constituent states of the empire should reflect that of its ruler. Recognizing the failure of his Great Plan, Charles V abdicated. His brother, Ferdinand (1556–64), assumed the role of emperor, while his son, Philip (1556–98), became king of Spain. Although the council had been adjourned, Julius III sought to champion the reforms developed at Trent. He appointed a commission drawing on reformers from both wings of the Church, including Carafa and Cervini on the one hand and Pole, Toledo, and Morone on the other. Together they produced a draft bull, *Varietas temporum*, based on the Tridentine decrees, which sought to control episcopal pluralism and enforce residence.[5]

The Inquisition of Heresy during the Pontificate of Julius III

The leading figure in the Roman Inquisition, Cardinal Carafa, had been opposed to the Council of Trent's resumption in 1551. Although he supported the principle of episcopal reform, he had reservations about these decisions being made at a council. He was also concerned about evangelical influence within the Catholic hierarchy and especially feared the continued influence of the *spirituali* on the fathers' discussion of doctrine. In the years that followed his attack on Pole in the conclave, he began to consolidate the Roman Inquisition's power and to use it to investigate high-ranking Catholic evangelicals. As Massimo Firpo has argued, this was part of a coordinated effort to prevent the *spirituali* from taking control of the Church. Julius did on occasion seek to constrain Carafa's power. For instance, in 1551 he intervened in the Roman Inquisition's efforts to prosecute the bishop of Bergamo, Vittore Soranzo (1500–1558). His actions may have been prompted by a desire to protect Cardinals Pole and Morone from being implicated by these investigations. Infuriated, Carafa encouraged the Roman Inquisition to secretly continue its investigations into the beliefs of Pole, Morone, and other Valdensians. His intention was to gather evidence of their heresy, which could, if necessary, be used against them.[6]

It is difficult to draw an accurate picture of the Roman Inquisition's activities during the 1550s. Although there is abundant evidence to suggest that the officials at the heart of the Roman Inquisition were preoccupied with the threat of Protestantism in its various forms, the picture from the outlying tribunals is less clear-cut. This is partly a reflection of the source material available to historians. While the archives of the Spanish Inquisition are largely intact, offering the possibility of accurately reconstructing its activities, the archives of the Roman Inquisition and its various tribunals are incomplete and widely dispersed. Of the various tribunals of Italy, the Venetian Inquisition has the best-preserved archives, and so it has been the focus of extensive research. The data from this tribunal shows that Protestantism was indeed the major focus of its activity during the first forty years of its existence. Of a total of 1,129 cases heard between 1547 and 1585, 707 cases concerned charges of Lutheranism alone, and a further 37 concerned charges of Anabaptism and 13 charges of Calvinism. As a point of comparison, there were only 59 prosecutions for magic during the same period.[7]

The predominance of Venice in the historiography of the inquisition of heresy in Italy may, however, have created some misapprehensions. To take one example, Matteo Duni cited the evidence from Venice, supplemented by data from Modena, to argue that pressure from the central Holy Office encouraged inquisitors to concentrate on the prosecution of Protestantism in the period circa 1542 to circa 1580. "As a direct result," he continued, "repression of magic gradually shifted from constituting a priority of inquisitors to becoming a merely residual function of their office."[8] While Duni is undoubtedly correct to suggest that investigations of Protestantism were more frequent than those of magic in Venice and Modena, this was not the case across all tribunals. For instance, between 1557 and 1595 the Friulian tribunal heard only three cases of people charged with "apostasy from the faith"—a charge embracing Lutheranism, Anabaptism, Calvinism, Orthodox Christianity, and Islam—but prosecuted sixty-two people for practicing the magical arts.[9] The pattern is similar in the kingdom of Naples, where, as we have seen, central control over inquisition was looser. Between 1564 and 1590, Neapolitan inquisitors investigated 19 people for Protestantism but 178 for practicing the magical arts. These examples suggest that if the desire to repress Protestantism triggered (or at least coincided with) a relative decline in prosecutions for magic in cities such as Venice and Modena, this was not necessarily true across Italy as a whole. The disparity between the rates of prosecution in these various case studies may reflect the circumstances faced by each tribunal. The relatively high number of Protestants or philo-Protestants in Venice and Modena may explain why Protestantism occupied so much of the local inquisitors' time and attention.[10]

Without further data from other tribunals, it is equally difficult to draw accurate comparisons between the number of people tried by inquisitorial tribunals for practicing magic before and after the establishment of the Roman Inquisition. It is, however, possible to provide some indications. Del Col estimated that between 1500 and 1541 in all Italian tribunals, whether secular or ecclesiastical, around five hundred people were charged with magical offenses. Currently, we possess data for only a limited number of tribunals directly accountable to the Holy Office in Rome—namely, Venice, Modena, and Friuli. They show that 154 people were accused of practicing some form of magic in an inquisitorial court in the period circa 1542 to circa 1590. While this figure is clearly lower than Del Col's estimate for the preceding period, it excludes the activities of numerous tribunals of the Roman Inquisition in central and northern Italy. The picture for Italy as a whole is further changed if we consider evidence from tribunals loosely controlled by the Holy Office. If we include data from trials conducted by tribunals active in the kingdom of Naples, for example, we arrive at a total of 333 accusations. It is also important to note that the branches of the Spanish Inquisition active in Italy saw a marked increase in their prosecution of magic during this period. In Sicily, for example, inquisitors investigated 177 people for these offenses in 1551–1600. This means that from a limited sampling of inquisitorial tribunals, and without including trials heard in either secular tribunals or episcopal courts, there were over five hundred trials for magic or witchcraft across Italy in the period 1542–1600.[11]

Without further data from other tribunals, any conclusions drawn from these statistics must be provisional. It is, however, possible to suggest that in the period circa 1540 to circa 1600 the total number of people investigated on charges of magic in Italy as a whole was at least comparable to that in earlier periods. The current lack of evidence from central and northern Italy means that it remains unclear whether there was a relative decline in the inquisition of magic in these areas. There were, however, important changes in the manner in which magic and witchcraft were investigated. As we have seen in chapter 6, mass trials for diabolic witchcraft centered on specific locations appear to have ended by the mid-1520s. The fact that the total number of individuals placed on trial for magical offenses remained constant indicates that although fewer people were investigated in any single trial, trials were becoming more numerous. The evidence also suggests that the inquisition of magic was spreading into new geographical areas. A comparison between Del Col's data on the inquisition of magic in the early sixteenth century and Tedeschi and Monter's studies of the later period shows that the prosecution of magic in the kingdom of Naples was extremely rare before the 1540s but that it escalated rapidly after the creation of the new inquisitorial structures.[12] This conclusion may, of course, simply be an illusion produced by the survival of sources or by the current state of research into the inquisition of magic in the kingdom of Naples before circa 1540. Combined with the evidence from Sicily, however, the data indicates that a new effort to eradicate magic began in southern Italy long before the effective suppression of Protestantism.

The Mendicants and Censorship

The preceding analysis suggests that although its relative importance varied across local tribunals, the pursuit of magic remained an integral part of inquisitors' activities during the period 1542 to circa 1580. The impetus to prosecute these arts came from neither the Council of Trent, which as we have seen paid no attention to such matters, nor Carafa and his associates, who were focused on the eradication of Protestantism. To explain this sustained interest in the pursuit of superstitious magic, it is essential to consider the ideas and practices of sixteenth-century inquisitors in the context of those of their predecessors, as Ruth Martin argued over thirty years ago.[13] The mendicant orders, especially the Dominicans, continued to play significant roles both in local tribunals and in the newly founded structures of the Roman Inquisition. The creation of the new institution afforded them a new means to impose on society standards of orthodoxy formulated within their orders. This was the most significant effect of the establishment of the Roman Inquisition on the prosecution of magic: for the first time, the friars could attempt to assert that their definition of the boundaries between orthodox and superstitious practices was the only acceptable one and in turn to impose it on Christian society as a whole.

The clearest evidence for the mendicants' efforts to harness the centralized structures of censorship to impose their conception of orthodoxy comes from book censorship. The Roman Inquisition's authority in this area expanded on April 29, 1550, when Julius III promulgated a bull that revoked all previously issued licenses to read or possess forbidden books and determined that henceforth only inquisitors and *commissari* could issue them.[14] These new powers were complemented by the Roman Inquisition's decision to produce an Index of Forbidden Books to guide the work of censorship. The idea of an index was not new. During the 1540s, the Universities of Paris and Louvain and several Italian states including Milan, Lucca, Siena, and Venice had drawn up lists of banned works. They were primarily concerned with works of theology, however. The indices prepared by the Italian states condemned no works of magic or divination. Those drawn up by the Universities of Paris and Louvain banned several specific texts—for instance, Henricus Cornelius Agrippa's *De vanitate scientiarum* and *De occulta philosophia* (Paris Index of 1544 and Louvain Index of 1546) and Girolamo Cardano's *De subtilitate* (Paris Index of 1551)—but they did not systematically tackle superstition.[15]

Work on the Roman Inquisition's index began in the early 1550s, coordinated by the Observant Dominican and inquisitor Michele Ghislieri (1504–72). He initially planned to print the most recent editions of the indices compiled at Louvain (1546) and Paris (1547). Quickly realizing that this measure would not suffice, he appointed two of his confrères to the task of compiling a new list. Although the resulting index was not officially promulgated, editions were printed in Florence, Venice, and

Milan.[16] The Roman Inquisition's list was made up of 596 individual entries. Many of these prohibitions could be seen as implementing—or at least complementing—Tridentine decrees. Like the earlier indices produced in Paris, Louvain, and the Italian states, it focused on restricting the circulation of the ideas of northern Reformers such as Martin Luther and John Calvin. Following earlier precedents, the index also banned the works of exiled evangelical authors such as Ochino and Vermigli and those of members of the *spirituali* such as the *Beneficio di Christo* that explicitly drew on the writings of northern Reformers. Indicating a desire to control Italian evangelical thought, the new index included previously unmentioned authors such as Juan de Valdés and Pier Paolo Vergerio.[17]

There are, however, indications that the Roman Inquisition was not only seeking to catalogue works that had been defined heretical by conciliar decree but also attempting to use this list to define orthodoxy. As we have seen, Carafa and his associates used the Inquisition to combat the *spirituali*, their perceived enemies within the Church. They were also concerned by various forms of evangelical Catholic thought. Their treatment of Erasmus in the new index is instructive. For many, he was an orthodox Catholic. Not only had he renounced the principle of justification *sola fide* and reaffirmed papal authority, but he had been feted by Charles V, and his ideas had inspired the early Jesuit order. He nevertheless remained a controversial figure within the Church, whose ideas were associated with the evangelical movement and, in Italy, the beliefs of the *spirituali*. His opponents in the Church had previously used the means at their disposal to curtail his influence. As we have seen, during the 1520s members of the Dominican order had used the Spanish Inquisition as part of their efforts to suppress Erasmus's thought and evangelism more broadly. In a similar manner, the Roman Inquisition's 1554 Index included nine of Erasmus's works. This decision exemplified how the members of the Roman Inquisition were beginning to use the new central structures of the Church to draw the boundaries of orthodoxy in ways not envisaged at Trent.[18]

There is also evidence to suggest that Dominican censors were using the central structures of censorship to promote a distinctive conception of orthodoxy constructed and maintained within their order. As we have seen, eradicating heresy had been central to the Dominicans' ministry since their foundation in the thirteenth century. Over the course of three centuries, the Dominicans, alongside the other mendicant orders, had generated a rich repertoire of writings dealing with heresy, ranging from theological treatises to sermons and inquisitorial handbooks. There is direct evidence to suggest that the compilers of the 1554 Index drew on these intellectual resources. Perhaps most obviously, they made explicit use of

Nicholas Eymerich's *Directorium inquisitorum* to guide their selection of prohibited works. By using these resources, the authors of the index infused it with a distinctively Dominican conception of the boundaries of orthodoxy, which did not necessarily reflect the opinions maintained in the Church as a whole.

Dominican influence over the 1554 Index can be demonstrated by its compilers' attitudes toward long-dead authors. Some of their decisions—for example, banning the writings of Jan Hus and John Wycliffe—were unexceptional. Each man had been condemned by a council, and their works had also been prohibited in earlier Italian lists. Other prohibitions clearly indicate that the Dominican compilers of this index were drawing on a section of Eymerich's *Directorium* (2.23–29) in which he reproduced a series of lists of previously banned works. They included thirteen texts written by Arnald of Villanova, although none concerned with magic. More controversially, following Eymerich, the authors of the 1554 Index also banned twenty works by Raymond Lull. As we have seen, Eymerich had played a leading role in attempts to condemn Lull as a heretic. Notably, he had included in his inquisitorial handbook a bull supposedly issued by Gregory XI, *Conservationi puritatis,* which formed the basis for the prohibition contained in the 1554 Index. Although the authors of the 1554 Index believed that this bull provided a basis for the condemnation of Lull's work, its validity was contested. By as early as 1419, the papacy had determined that it was spurious. Although Dominicans subsequently questioned Lull's orthodoxy, his ideas were openly taught and discussed, especially in Spain. The inclusion of his works in an index issued under the auspices of the papacy offered the Dominicans a means to establish as definitive their contested assessment of his writings.[19]

The evidence from the 1554 Index also suggests that its Dominican compilers were not simply concerned by heresy but sought to use the list to achieve further ambitions that were consistent with the program of pastoral reform established in their order. For instance, they continued the mendicants' attacks on Italy's Jewish population by prohibiting the Talmud.[20] Hostility to the Jews was, of course, not the exclusive preserve of the mendicants. Efforts to regulate the Jews' presence in Christian society was embedded in various reformist traditions. Earlier councils, notably Lateran III and IV, had codified rules determining the terms of their existence in Christendom, while more recently Lateran V had sought to address the problem of Judaizers and endorsed the use of *montes di pietatis.*[21] Furthermore, a number of sixteenth-century reformers, including Giustiniani and Querini in their *Libellus ad Leonem X,* had articulated the desire to encourage Jewish conversion to Christianity. None of these measures were specifically concerned with the Talmud, however, and neither this text nor the Jews were mentioned at Trent. The specific decision to target the Talmud appears to have been driven by the friars. They had been fomenting anti-Judaic sentiment since the thirteenth century, and these themes had been a consistent feature of their preaching in Italy throughout the late fifteenth and early sixteenth centuries. The friars had also frequently been involved in attacks on the Talmud, sometimes while acting in the capacity of inquisitors. These actions continued into the sixteenth century and gained strength after the foundation of the Roman Inquisition. Indeed, by 1553—that is, at the same time as the first inquisitorial index was being prepared—the Roman Inquisition was playing a leading role in attempts to suppress the circulation of the Talmud in the cities of Rome and Venice.[22]

The pursuit of superstition also remained a central part of the mendicants' program of pastoral reform. Once again, they used the index to pursue these ambitions. The compilers of the 1554 Index banned several specific texts describing magical and divinatory arts and implemented general bans against certain categories of superstitious arts. Simply including general prohibitions on works containing magical and divinatory arts was an innovative use of the index, but its Dominican compilers used the list to define the boundaries of superstition. Many of the key ideas were consistent with existing inquisitorial practice, as codified by Eymerich in his *Directorium inquisitorum*. Perhaps the most obvious example of this approach in practice was their decision to ban works describing "necromancy." This was a broad category that designated forms of ritual magic that had been of intense concern to Eymerich. It included the use of magical arts that explicitly invoked demons and those used ostensibly to solicit the assistance of angels. It is therefore no surprise that the index also specifically banned specific theurgic texts such as the *Ars notoria*.[23]

The 1554 Index's ban on necromantic texts indicates important continuities with earlier inquisitorial theory and practice, but this picture is complicated by a series of prohibitions on various divinatory arts. Although the list made no reference to predictive astrology, it banned works of geomancy, hydromancy, and pyromancy. Reinforcing these general bans, the index also proscribed the writings of Bartolomeo della Rocca Cocles, an author who had produced texts describing predictive arts such as chiromancy.[24] These prohibitions point to a significant change in the criteria used by the compilers of this index to determine the boundaries between orthodox and superstitious operative arts. As we have seen in previous chapters, Eymerich had implied that it was possible to practice any art that contained no explicit ritual element. He therefore concluded that chiromancy, for example, was legitimate.

At face value, the decision made by the authors of the 1554 Index to ban these arts could be interpreted as a rejection of Eymerich's authority. As we have seen, however, the compilers of this index drew on a list of banned books provided in part 2 of the *Directorium*. In addition to bans on the writings of Arnald of Villanova and Lull, these prohibitions included a decree issued at an unspecified date in Paris that proscribed several operative arts including the ostensibly natural arts of physiognomy and chiromancy.[25] Eymerich had therefore provided two potentially contradictory means of determining the legitimate boundaries of divination, each underpinned by a differing rationale. Inquisitors had clearly read this section of the *Directorium*, for they reproduced this specific ban and several others as an appendix to the 1554 Index. A later document, *Instructiones nonnulle circa libros nominatim prohibitos in Santo Indice* (Instructions regarding books named as prohibited in the Holy Index), dated to 1557–58, also cited this prohibition and indicated that it should be used as the basis for determining the legitimacy of the divinatory arts. By promoting this prohibition banning books that described ostensibly natural arts, the authors of this index had embraced censorial criteria that contradicted those described by Eymerich in the body of his handbook and subsequently used in established inquisitorial practice.[26]

To explain why the compilers of the 1554 Index based their definition of divination on this reading of Eymerich's handbook, it is necessary to recall the history of the interpretation of Thomist thought recounted in earlier chapters of this book. In *Summa theologica*, Aquinas had argued that the only way to determine the legitimacy of an art that contained no necessary ritual element was to determine whether it functioned by natural means. In other words, inquisitors were obliged to establish the limits of the natural order. Eymerich's definition of magic and divination was based on an idiosyncratic reading of Aquinas's ideas, which had led him to accept the orthodoxy of certain operative arts such as chiromancy that his illustrious forebear had condemned. Over the course of the fifteenth century, a number of Dominican friars, notably theorists of magic such as Johannes Nider and Henricus Institoris and Jacobus Sprenger, had established a more accurate reading of Aquinas's thought. While these ideas had influenced some members of the intellectual elite of the Church in the later fifteenth and early sixteenth centuries, they do not appear to have permeated down to practicing inquisitors. The construction of a centralized organization charged with the task of determining the boundaries of orthodoxy appears to have triggered a reckoning with the new works of witchcraft and magic. It enabled Dominican friars to embed the new interpretation of Aquinas's ideas that had been developing from the fifteenth century into the Roman Inquisition's definition of orthodoxy. This conception of Thomist thought did not entirely displace Eymerich's interpretation of Aquinas's ideas, however. Influential figures in the apparatus of censorship continued to defend Eymerich's proposed means to define the boundaries of superstition, a situation that created the potential for disputes about the legitimacy of specific arts.[27]

Paul IV and the First Papal Index

The promulgation of the 1554 Index was a significant moment in the history of the sixteenth-century Church. It demonstrated Carafa's ambition to use the Roman Inquisition as a tool to root out not only Protestantism but also forms of Catholicism that he considered heterodox. As we have seen, its contents also indicated that members of the mendicant orders—above all the Dominicans—were using the Roman Inquisition as a vehicle to promote their own conception of Christian society and to fulfill long-standing ambitions such as controlling the circulation of Jewish writings and combating superstition. In effect, they had set out an alternative to the program of reform articulated at Trent. Although the members of the Roman Inquisition were powerful in the mid-1550s, their ability to impose their vision of orthodoxy on Italian society was held in check by a watchful pontiff and a skeptical episcopate, while their ability to act was hampered by weak local organization. This situation was changed by a series of contingent events that began with the death of Julius III on March 23, 1555. He was succeeded by another former Tridentine legate, Cardinal Cervini. The new pontiff, a reformer with traditionalist doctrinal views, took the name Marcellus II. He died less than a month later. In the subsequent conclave, Carafa was elected as Paul IV on May 23, 1555.[28]

The new pontiff was deeply opposed to any attempts to negotiate with the Protestants, and he had been enraged by the Habsburgs' compromises with the Lutherans at the Peace of Augsburg. While he remained a committed reformer, he believed that change should be directed by the papacy. Consequently, he not only rejected any efforts to reconvoke the Council of Trent but also sought to suppress the reforms described in *Varietas temporum* that were largely modeled on Tridentine decrees. To formulate an alternative program of reform, he set up a commission composed of sixty prelates. The commission subsequently proposed reforms based on the *Consilium* of 1537, which were designed to revitalize episcopal structures of pastoral supervision and discipline first developed in the late twelfth and early thirteenth centuries. In this respect, Paul IV's proposals were consistent with reformist aspirations articulated at Trent and therefore also built on proposals made at Lateran V. Yet while Paul's reforms recognized the importance of the episcopate, his actions were highly provocative. Rejecting a council necessitated imposing changes on bishops rather than engaging them in the decision-making process. Such an approach was a significant affront to an episcopate that had been asserting its collective authority throughout the fifteenth century, and it was a direct challenge to the ecclesiology of a number of those bishops who argued for divine origins of their office.[29]

Paul combined his reform agenda with rigorous attempts to root out heresy and corruption from within the Church and Christian society at large. Once more, his methods threatened to set him at odds with the episcopate. He sought to place the Roman Inquisition at the center of a disciplinary system that would owe its allegiance directly to the papacy, a move that enabled him—if necessary—to circumvent the independent action of the local bishops. To consolidate the Holy Office's position within the curia, Paul identified it as the most important of the Congregations of Cardinals. He also extended his use of the Roman Inquisition as a means to define the boundaries of orthodoxy. On May 31, 1557, Cardinal Morone was arrested on Paul's orders, charged with holding Lutheran beliefs, and imprisoned in the Castel St. Angelo. Pietro Carnesecchi, who had been associated with both Juan de Valdés's circle and Pole's group in Viterbo and who was then resident in Venice, was condemned in absentia as a heretic. In a separate action, Paul IV stripped Cardinal Pole of his status as legate to the new queen of England, the Catholic Mary I (1553–58).[30]

In 1556, Paul IV appointed a commission to prepare a new papal index. Its members included Ghislieri, two Augustinian friars, and the general of the Jesuit order, Diego Lainez. Reflecting Paul IV's enhanced ambition, the new list was intended to apply not simply in individual cities but throughout Italy and indeed the wider Catholic world. Work on the new list was completed by November 1557, and it was printed the following month.[31] The compilers of this list used it to pursue many of the objectives established in the 1554 Index. Most obviously, they continued the earlier efforts to restrict access to the writings of northern Reformers and those of Italian reformers such as Ochino, Vermigli, and Vergerio. The compilers of the 1557 Index also continued and in many respects expanded the controversial reforming agenda established in the earlier index. In an intensification of Paul IV and the Inquisition's efforts to suppress Catholic evangelism, they prohibited the complete works of Erasmus.

The compilers of the 1557 Index also continued the efforts to use the index to pursue a wider program of social discipline rooted in the mendicants' program of pastoral reform. For example, they again banned the Talmud and any glosses, annotations, or expositions of this text. These actions were supported by papal decrees that confined the Jews to ghettos and mandated that they wear identifying badges.[32] Continuing the Roman Inquisition's campaign against superstition, the authors of the 1557 Index once more introduced a general prohibition against necromancy and bans on the *Ars notoria* and the *Clavicula salomonis*. A further general proscription banned "all books and writings of the magical disciplines." They also extended the ban on Arnald of Villanova's writings to include all his works, including those that touched on magic, and not simply those identified by Eymerich. They prohibited too, several contemporary works such as Johann Reuchlin's *De arte concionandi* and the complete works of Cornelius Agrippa.[33] Finally, the authors of the 1557 Index refined the criteria for censoring divination. In addition to banning several individual arts, they drafted a general injunction that proscribed

> all books and writings of chiromancy, geomancy, hydromancy, physiognomy, pyromancy or necromancy, or those in which they are written down sortilege, sorcery, incantations, magical divinations, or astrological judgements, regarding horoscopes, nativities, future events, or regarding the particular outcomes, circumstances, of the life or death of any man.[34]

While this prohibition included many of the arts banned in the 1554 Index, it extended its scope to include astrology for the first time.

The scope of the 1557 Index's ban on divination suggests that its compilers had once more drawn on the criteria established by Aquinas in his *Summa theologica* to identify illicit forms of divination. The manner in which they dealt with astrology suggests that their engagement with his legacy was in fact complex. Unlike Aquinas, the compilers of this index did not explicitly distinguish between different types of astrological practice, nor did they draw any distinctions between this art and any other form of divinatory activity. In fact, the 1557 Index's explicit ban on works describing arts that promised to produce privileged knowledge of future events in the life of an individual appeared to question the legitimacy of several astrological practices that Aquinas had considered orthodox. For example, its wording made it hard if not impossible to defend the idea that the stars could incline an individual toward a particular action. Furthermore, instances of the failure of health were examples of future events affecting the circumstances of the life and sometimes the death of an individual.

The wording of the 1557 Index therefore also appeared to refute Aquinas's contention that the stars could be used to predict the likely, though uncertain, future health of an individual or a population. With the parameters of astrology thus constructed, it seemed that its practitioners could only make predictions relating to such events as the timing of eclipses.

If this reading of the 1557 Index's prohibition of astrology is correct, it suggests that its authors did not seek to establish Aquinas's understanding of the orthodox parameters for astrology as the standard for censorship. Instead, it seems that they were articulating a conception of astrology far closer to that of earlier ecclesiastical authorities such as Augustine, who had condemned astrological prediction tout court. How can this be explained? As we saw in chapter 4, prompted by Pico della Mirandola's *Disputationes adversus astrologiam divinatricem,* Christians had begun to question established assumptions about astrological knowledge. This epistemological crisis affected Dominican reformers such as Girolamo Savonarola, who had appeared to develop doubts about Aquinas's views on astrology. Further research is necessary, but it may be that as a result of this growing debate over the status and validity of astrology, the compilers of the 1557 Index in fact rejected key aspects of Aquinas's definition of natural astrology.[35]

For reasons that are unclear, the 1557 Index was never promulgated. Members of the Roman Inquisition appear to have accepted that it was in some respects unsatisfactory, for they soon began to work on a revised list. They completed this task in late 1558 and promulgated the new index the following January. Reflecting Paul IV's desire to centralize disciplinary authority, he announced that the new list was to be implemented solely by inquisitors, thereby—in principle at least—removing local bishops from the censorial process. On January 5, 1559, Paul IV issued a further bull directed at confessors. It reminded them that they could not absolve anyone who adhered to heresies and also required them to ask penitents whether they possessed any forbidden books, whether they had read them, and whether they knew of others who had. If a penitent confessed to any of these crimes, the bull stated that the priest was obliged to send them to either the Holy Office or their local bishop. Should the confessor fail to comply, they would be subject to the penalty of *scomunica maggiore,* "reserved only" to the pope or inquisitor general. These instructions were designed to enhance central oversight of the Church's disciplinary processes, but, as contemporaries recognized, they held significant implications for the secrecy of confession.[36]

The 1559 Index was far longer than the Roman Inquisition's earlier lists, running to a total of 1,168 entries. In several important respects, however, it was consistent with established principles of censorship. It contained several general bans, which tended to reinforce rules governing the printing industry that had been gradually introduced since the late fifteenth century by papal bulls and decrees passed at Lateran V. For example, it prohibited several specific classes of books, including all works previously condemned by either popes or councils before 1515, all books published in the preceding forty years that lacked titles, and those without an imprimatur issued by either an inquisitor or a bishop. Like its predecessors, this index also proscribed a series of specific books and authors. The overwhelming majority of these works were theological texts produced during the sixteenth-century religious crisis, including a significant number that contained ideas determined to be heretical by the Council of Trent. Foremost among them were the writings of leading Protestant Reformers such as Luther, Melanchthon, and Calvin.[37]

The compilers of the 1559 Index may have built on measures established at earlier councils, including Trent, but they once again used this list to set out an independent conception of orthodoxy. For example, the 1559 Index prohibited the complete works of evangelical authors such as Erasmus. Its authors also established measures to control access to scripture by banning Catholics from possessing or reading vernacular Bibles, if they had not first obtained a license issued by the Holy Office. This set of prohibitions went far beyond the position established at Trent, which, as we have seen, had not expressed a clear opinion on the orthodoxy of vernacular scripture. Furthermore, by giving inquisitors sole responsibility for issuing licenses to read scripture, the compilers of the 1559 Index prevented bishops from having any discretion in these matters. The compilers of this index reinforced their efforts to control reading practices by implementing bans on several editions of the Bible and unauthorized versions of the New Testament. They also proscribed owning or reading editions of the church fathers or other ancient authors that had been translated or annotated by a heretic. Uniquely, this index also included the names of fifty-nine printers who were held responsible for the publication of heretical material.[38]

The compilers of the 1559 Index continued to use the list to promote conceptions of orthodoxy that were consistent with the Dominicans' long-standing reform program. The index once again placed a ban on Raymond Lull's writings. It also prohibited the Talmud and its glosses, annotations, or interpretations. In addition to these measures, the compilers of this index continued their efforts to control books and writings that described the magical and divinatory arts. The 1559 Index contained a general prohibition against necromantic works and a ban on all books and writings on the magical arts. Reinforcing these efforts, the Index continued to proscribe individual texts such as the *Clavicula salomonis* and the *Ars notoria*. This index also banned the complete works of Agrippa and three works by Reuchlin, including his *Ars cabalistica*.[39] Like the indices of 1554 and 1557, the 1559 Index also banned various individual arts and included a general prohibition, which proscribed

> all books and writings of chiromancy, physiognomy, aeromancy, geomancy, hydromancy, onomancy, pyromancy or necromancy, or those books and writings in which they are written down sortilege, sorcery, augury, haruspicy, incantations, divinations from the magic arts or judicial astrology regarding future contingent events or the results of events or cases of chance, having excepted all those natural observations, which are useful to the arts of navigation, agriculture or medicine.[40]

The wording of this general prohibition not only reiterated the ban on ostensibly natural arts such as chiromancy, established in previous indices, but also maintained the effort begun in the 1557 Index to define the appropriate boundaries of astrology.

Although the 1559 Index's general prohibition relating to divinatory arts was similar to the one contained in the 1557 Index, its authors took a more nuanced approach to astrological prediction. The wording of the 1559 Index specifically permitted astrologers to use techniques that made knowledge by "natural" means, which were useful to medicine, astrology, or agriculture. By using such language, the compilers of the 1559 Index appeared to be invoking the Thomist principle that an effect could be predicted if it had a single, observable, natural cause. As we have seen, this idea could be used to justify a wide range of astrological practices, including predicting likely (albeit uncertain) future human actions. The compilers of this index nevertheless only accepted the use of astrological prediction in fields such as medicine. It is possible, as Ugo Baldini has suggested, that they had intended the process of censorship to be guided by the principles that Aquinas had established in the *Summa theologica*, including belief in the idea of astral inclination, which justified the practice of making uncertain astrological predictions about future human actions.[41] I suggest, however, that although there is evidence that the compilers of this index drew on Aquinas's ideas to justify some forms of astrological prediction, we cannot assume that they had intended to implement them in full. The principles governing the censorship of astrology in 1559 can perhaps be best described as a limited or variant form of Thomism.

The compilers of the 1559 Index introduced a further element of ambiguity into the practice of censorship when they addressed, or at least appeared to address, the orthodoxy of astrological talismans. While the compilers of earlier inquisitorial indices had sought to prohibit various forms of divination, including astrology, their efforts to control texts describing astrological talismans were at most limited. The new index indicates that by the late 1550s members of the Roman Inquisition were beginning to consider their legitimacy. To take an important example, the compilers of this index banned Pietro d'Abano's *Liber de imaginibus astrologicis, et de omni genere divinationis* (*Book of Astrological Images and of Every Type of Divination*), which Jesús Martínez de Bujanda has identified as the *Conciliator*.[42] From the extant documentation, it is unclear whether the compilers of this index objected to the *Conciliator*'s description of divinatory practices or the production of astrological images or, indeed, both. As we have seen, the legitimacy of astrological images had been disputed within the Dominican order since the thirteenth century, and leading authorities such as Albertus Magnus and Aquinas had held conflicting views on this question. By the fifteenth century, however, Dominican friars such as Nider, Institoris, and Sprenger had condemned their production on Aquinas's authority. While it might therefore be expected that the Roman Inquisition would ban texts describing such practices, its members appear uncertain about their legitimacy. For example, although Marsilio Ficino's *De triplica vita*, one of the most significant contemporary works treating astrological images, was not included in this or indeed any later index, inquisitors continued to scrutinize this work. They also later questioned the orthodoxy of other texts such as Arnald of Villanova's *Antidotarium*, which described the production of astrological images.[43]

In the 1559 Index, Paul IV and the Roman Inquisition had articulated an ambitious program designed to reshape the acceptable parameters of Catholic doctrine and practice, which was backed by an audacious attempt to centrally coordinate its implementation. In practice, numerous obstacles prevented their vision of reform from being realized in full. The document *Instructio circa indicem*, issued by the Holy Office shortly after the index's promulgation in January 1559, revealed the distance between the Inquisition's ambition and its actual power by drawing attention to the fact that its local organization remained weak and ineffectual.[44] The Roman Inquisition's original intention was to directly control the process of censorship by stating that the new list was to be enforced only by inquisitors. They or their appointees were to have sole responsibility for the seizure of books. The *Instructio* tacitly acknowledged the limitations of the Roman Inquisition's power by stating that when an inquisitor was not present, the ordinary could implement the index. It also indicated that the Holy Office recognized that the 1559 Index contained several ambiguities relating to the standards to be used in censorship. For example, the *Instructio* clarified the rules for censoring the art of physiognomy. It noted that although books describing this art were prohibited in the index, censors should only ban texts that described how to use it for purposes of divination—for example, those that described the practice of chiromancy. This advice was also ambiguous, however. Although consistent with Aquinas's views, it did not consider the possibility suggested by Eymerich that chiromancy might also be considered a natural art.[45]

The Third Period of Trent

Paul IV died in August 1559. In Rome, news of this event was marked by rioting. While disorder was common during the *sede vacante*—the period between the death of one pontiff and the election of his successor—the violence of that summer was unusual in both its extent and its severity. In addition to destroying statues of Paul IV, the crowds turned their attention to another monument to his pontificate: the Palace of the Inquisition. A mob ransacked the building, burning records and documents and freeing its prisoners. The conclave to elect Paul's successor began on September 5 and concluded on December 25 with the election of Milanese cardinal Giovanni Angelo de' Medici (1499–1565). It soon became clear that the new pope, who took the name Pius IV, rejected key aspects of his predecessor's regime. Almost immediately, he ordered the release of Cardinal Morone from the prisons of the Castel St. Angelo. Despite this symbolic act, under the leadership of Michele Ghislieri, members of the Roman Inquisition continued their campaign against other former members of the *spirituali*. In 1560–61, they launched another trial against Pietro Carnesecchi. Although Carnesecchi was eventually released on this occasion, these actions show Ghislieri's ongoing desire to eliminate the threat posed by those he considered heretics.[46]

Pius IV also gave early indications that he would be willing to rescind a number of the measures that his predecessor had taken to centralize oversight of disciplinary processes, thereby returning authority to local bishops. Most notably, in late 1559 he formed a commission to review the index. The election of the new pope also rekindled hopes that the Council of Trent could be reconvened and concluded. Although Charles V's Great Plan had ended in ignominious failure, both the Spanish king Philip II and Emperor Ferdinand continued to support conciliar reform. Indeed, even at this late stage, Ferdinand hoped that reconciliation with the Lutherans might still be possible. They were also joined by Henry II of France, who had committed to supporting a council as part of the terms of the Treaty of

Cateau-Cambrésis that he had concluded with Spain in April 1559, which ended the damaging Habsburg-Valois wars. Pius was initially ambivalent about these proposals. During his conclave, he had given a conventional undertaking to convene a universal council, but once elected he gave little indication that he would follow through on his promise. His commitment to the idea of a council was transformed by events in France. The premature death of Henry II and the accession of his teenage son Francis II to the throne in September 1559 coincided with the outbreak of religious unrest. The French monarchy proposed to hold a national council. Fearing that this could lead to the establishment of a separate Church of France, Pius declared his intention to hold a universal council.[47]

The proposal to hold a universal council presented Pius IV with a series of problems. The rulers of Spain, the empire, and France disagreed as to whether this was to be a new council or a continuation of the old one. Since the decrees agreed on during the previous periods of the Council of Trent had not been formally ratified, some argued that they could be redefined. This was an attractive prospect for both Ferdinand and Francis, as they sought compromise with the Protestants in their states. Although open to negotiation with the Protestants, Pius doubted that a council could resolve the schism and conceived it as a forum to establish Catholic doctrine and practice.[48] He resolved the problem by issuing an ambiguous bull of convocation (November 29, 1560), which lifted the suspension of the earlier council while not explicitly describing the proposed meeting as a continuation of the old one. Pius also had to determine how he would deal with proposals for reform. Many bishops continued to advocate for changes that would enhance the role of the episcopate and place them at the heart of Catholic reform. Although Pius had sought to correct some of the excesses perpetrated during the reign of his predecessor, he was wary of allowing the council to pass decrees that could serve to undermine papal authority.[49]

Pius appointed five legates to run the council. They were led by Ercole Gonzaga, the "first president," and included Girolamo Seripando and the bishop of Pesaro, Ludovico Simonetta. The latter had been a member of the curia since 1549 and was a staunch advocate of papal authority; he would play an increasingly important role in the later stages of the council.[50] While waiting for the delegates to assemble, the legates began preparations for the council. They were especially eager to receive the results of the work undertaken by the commission charged with the reform of the 1559 Index, for they were concerned that the severity of the list might hamper any chances of reconciliation with the Protestants. In May 1561, Cardinal Carlo Borromeo, the papal nephew and secretary of state, had promised the legates that they would soon receive a revised list, but he instead sent them a further document, the *Moderatio indicis*, written by Ghislieri. Building on the earlier *Instructio circa indicem*, it proposed changes to some of the principles governing the operation of the index. The legates responded by writing to the pope requesting permission for the council to review the Inquisition's index. It was granted on January 14, 1562. When the council formally opened four days later, one of the delegates' first acts was to appoint a commission to prepare a draft decree concerning the index. The Tridentine fathers accepted the text of this decree during the eighteenth session, held on February 26. They subsequently appointed a commission of eighteen individuals to prepare a new index, the members of which were, in accordance with the fathers' wishes, drawn from different orders and from different nations within the Church.[51]

While the Index was being prepared, the council fathers continued their deliberations. Reform was at the top of the agenda. Seripando, one of the papal legates, appointed five Italian bishops to produce a memorandum on the matter, and they proposed a series of measures designed to reinvigorate the episcopate. Episcopal residence remained a pressing concern. Measures implemented in earlier sessions of the council that were designed to deal with this matter had been widely ignored, and many reform-minded bishops considered them a failure. One of the proposals made in the memorandum was—if all else failed—to declare that residence was required *de jure divino*, meaning that not even a pope could grant dispensation from this duty. This proposal threatened to reignite the disputes over this issue that had flared up during the first period of Trent. To avoid this potential conflict, Seripando submitted a reduced document to the council for consideration that omitted the contentious idea that residence was *jus divinum*. The majority of the non-Italian bishops, supported by some Italians—notably, Seripando's associate Ludiovico Beccadelli—nevertheless demanded the right to discuss this matter in the General Congregation. The ensuing dispute polarized the council. Proponents of residence *de jure divino* were opposed by a second group of fathers, consisting principally of Italian bishops and led by the legate Simonetta. They maintained that adopting this principle contradicted established Church practice, that it would harm the functioning of the curia, and that it would impinge on papal authority.[52]

The fathers' renewed disputes over the principle of residence *de jure divino* altered the balance of power between the legates. In his letters to Rome, Simonetta undermined the pope's trust in his fellow legates and expressed his concern that under Spanish influence the council was poised to declare its authority over the papacy. Influenced by Simonetta, Pius IV wrote to the legates admonishing them for their handling of the residence question and, in an indication of his failing trust in Gonzaga, threatening to send a further two legates. Borromeo wrote separately to Simonetta instructing him to ensure that the fathers did not discuss the question of whether residence was required by divine law. Pius IV subsequently ordered the postponement of discussion of residence altogether. This decision enraged the Spanish party, who responded by threatening to withdraw from the council. They were placated only by the legates' promise that the question of episcopal residence would indeed be discussed when the council considered holy orders.[53]

The council continued through the spring and early summer of 1562, but the two sessions held during this period accomplished nothing more than postponing the publication of further decrees. During Session 21, held on July 16, 1562, the fathers passed doctrinal decrees relating to the Eucharist, notably affirming that receiving the mass in both kinds was not necessary for salvation. They also accepted further limited reform decrees concerning episcopal benefices and administrative responsibilities. During the subsequent session, held on September 17, the council fathers returned to the question of whether communicants should be allowed to take communion in both kinds, a matter on which the fathers had delayed pronouncing in 1551. The fathers once more avoided making a decision and instead declared that the pope should pronounce on this issue at a future date. The fathers again passed a further limited reform decree. The final decrees set the date for the next session as November 12 and specified that the fathers would address the two remaining sacraments: holy orders and marriage. The legates hoped that once these last outstanding doctrinal questions were resolved, it would soon be possible to conclude the council.[54]

The Tridentine Index

In 1562, the Tridentine commission revised the 1559 Index. The resulting document echoed reforming bishops' desire to place the episcopate at the heart of efforts to define, promote, and—where necessary—enforce reform. Rejecting the centralizing tendencies of the Pauline era and seeking to rein in the Roman Inquisition's overweening power, the revised index restored the bishops' authority over printing and asserted that they should enjoy equal authority with inquisitors in matters of censorship. The commission also introduced a series of innovations, many of which were expressed in a series of rules governing various aspects of the practice of censorship. The new index was to be arranged around three classes of prohibited texts. The first was a list of authors whose works were entirely banned. The second class consisted of individual proscribed works. The final class was constituted of works that were suspended *donec corrigantur*—that is, "until corrected." The addition of this final category was a significant innovation: it not only introduced the principle that censors could remove offensive passages from otherwise acceptable works, thus allowing them to be returned to circulation, but also added the task of expurgation to the censors' role. The commission also used the new index to revise the mechanisms of censorship and redefine the respective roles of the inquisitors and bishops. Rule IV, for example, tacitly accepted the need to restrict access to scripture in the vernacular but nevertheless extended the power to issue licenses to read translated Bibles to both the episcopate and inquisitors.[55]

The commission also revised the content of the 1559 Index, reviewing existing entries and considering new ones. They continued to take a hard line on the writings of heresiarchs such as Luther and Calvin and maintained the complete prohibition of the writings of figures such as Juan de Valdés and Peter Martyr Vermigli. There were, however, vigorous objections to several of the bans imposed by the 1559 Index. The censorship of Lull's work quickly emerged as one point of contention. In a letter dated October 19, 1562, Camillo Campeggi, a Dominican and secretary to the commission, wrote to the inquisitor of Cremona in Rome. He recalled that some four months earlier, supporters of Lull had objected to his inclusion in the index. Campeggi indicated that Lull's supporters were attacking the basis for this decision—that is, a bull supposedly written by Gregory XI that Eymerich had included in his *Directorium*. Since they had been unable to find a copy of this bull, they believed that it was fake. Lull's supporters, Campeggi continued, also attacked Eymerich's character. Referring to his banishment from Aragon, they observed that as a punishment for his false accusations against Lull, Eymerich had been "deprived of the office and rank that he held." Campeggi concluded his letter with the observation that "they also say that Lull is regarded as a Saint in Spain and that he works miracles." Campeggi appeared unconcerned by these objections and remarked to Polizzi, "For us it will be enough if we can find that the *Directorium* had been approved by one of the Popes, and if the author had died with a reputation as a good man, therefore V.S., see if you can throw some light on these matters."[56]

The exchange between Campeggi and Polizzi demonstrated the process by which orthodoxy was being constructed within the Tridentine Church. Contemporaries understood that the Indices of Forbidden Books could be used to universalize ideas and principles originally formulated in localized contexts, whether geographical or institutional. One influential party, the Dominican order, had attempted to use the 1559 Index to unilaterally impose their assessment of Lull's orthodoxy on the Church as a whole. The index continued to offer a powerful instrument for defining orthodoxy, and both the Dominicans and Lull's supporters understood that in order to promote their ideas they needed to influence the Tridentine commission. As Campeggi's account indicated, however, in this instance the decision-making process depended less on the commission's examination of Lull's works than on their assessment of Eymerich's authority. He therefore urged Polizzi to find documents in Rome that could support their case. Lull's supporters, meanwhile, sought to undermine the validity of his inclusion in the index by attacking Eymerich's integrity and the credibility of the judgments that he had reached. These arguments not only threatened to undermine the basis for Lull's condemnation but also challenged the Dominicans' use of the *Directorium* as a basis to establish universally binding definitions of the boundaries between orthodoxy and heresy.

Subsequent correspondence between Campeggi and Polizzi reveals that this process continued through the autumn. In a letter dated November 16, 1562, Campeggi wrote, "With your letters of the 28 and 30 of the past month, I have received the condemnation of the Talmud produced by Innocent IV and also the testimony of Bernardus of Luxemburg against Lull."[57] Acknowledging his receipt of the work of Bernardus of Luxemburg (d. 1535), a Dominican friar critical of Lull's work, Campeggi revealed that the two men continued to search for documents to bolster the case against Lull. The preceding reference to Innocent IV's condemnation of the Talmud shows that they also used the same technique to influence the commission's work in other contested areas. Ultimately, however, the Dominicans' efforts were unsuccessful. The commission removed Lull from the 1564 Index and permitted the production and circulation of editions of the Talmud, on the condition that they did not bear this title and that any passages deemed offensive to Christians were removed. While this decision in principle allowed for the continued circulation of the Talmud, in practice it initiated a long series of disputes about how the text should be censored. In a further blow to the Inquisition, the new commission also took a more moderate approach to some evangelical authors. Notably, the new list relaxed the censorship of Erasmus. Only a selection of his works (including *Moriae encomium*) were now banned outright; his writings concerning religious matters were suspended pending revision and possible expurgation.[58]

The Tridentine commission also reviewed the 1559 Index's entries concerning superstitious arts. This was a significant intervention, because it was the only occasion that any cleric associated with the Council of Trent directly considered these matters. When dealing with superstition, the commission maintained the vigorous attempts made in the 1559 Index to separate licit from illicit forms of magic and divination. As we have seen, in earlier indices the criteria used to censor works describing superstitious arts almost certainly originated within the Dominican order. There were clear aspects of continuity between the work of the Tridentine commission and the Roman Inquisition's earlier program of censorship, for the new list appeared to base its censorship of the operative arts on the broad Thomist principles established in previous indices. It once again banned all works of necromancy and "Notoriae artis opera," and it prohibited specific texts such as the *Clavicula salomonis*. Agrippa was included in the first class of totally prohibited authors, although the works of Reuchlin, banned outright in the previous index, were selected for examination and expurgation.[59]

The 1564 Index also formalized previous prohibitions against works of divination and illicit magic, drawing them together in Rule IX of the index. It began by stating that "all books and writings concerned with Geomancy, Hydromancy, Aeromancy, Pyromancy, Onomancy, Chiromancy, Necromancy, or in which there are contained Sortilege, Veneficia, Augury, haruspicy, incantations of the magic arts, are completely rejected."[60] The only significant innovation in this portion of the rule was the omission of physiognomy from the list of totally prohibited arts, a change that drew the Tridentine Index into line with the censorial practices established in the *Instructio* of 1559. Rule IX continued, however, by redefining the boundaries of orthodox astrology. It declared that

> the bishops, however, shall see to it diligently that they should neither be read nor possessed, books, treatises or indices of judicial astrology, which, with regard to future occurring outcomes, chance events, or those actions which depend on the human will, dare to affirm anything that is *certain* to occur. They are permitted, however, judgements and natural observations, which are composed in order to be of use to the arts of navigation, agriculture or medicine.[61]

The wording of this passage is significant in two key respects. In the first instance, it appears to assign responsibility for determining the orthodoxy of astrological texts and writings to the bishops alone. The Tridentine commission charged with preparing the index was, it seems, now seeking to exclude inquisitors from key aspects of censorial work. Second, its members appeared to propose a definition of orthodox astrology that subtly altered the principles of censorship established in the 1559 Index.[62]

Since it allowed natural astrological predictions useful to medicine, agriculture, and navigation, the definition of orthodox astrology offered in the Tridentine Index was broadly consistent with the one put forward in 1559 Index. Rule IX added an important change, however. It prohibited astrologers from predicting events that were "certain" to occur. The implications of the passage were unclear to contemporaries, and they have subsequently been debated by historians. While it seemed to proscribe the practice of making *certain* predictions about future contingent events, it nevertheless suggested that it was possible to offer *uncertain* predictions about them. This position therefore clarified the legitimate use of astrological prediction in activities such as medicine. It confirmed, for example, that it would be possible for an astrologer to predict a likely future contingent event such as the outbreak of pestilence, but only if he did not assert that this event would necessarily come to pass. The inclusion of the word *certain* also introduced an element of ambiguity into the interpretation of the rule. Some contemporaries believed that it was possible to interpret it in such a manner that it legitimated the use of astrology to predict uncertain human actions—that is, they believed that it permitted the use of the doctrine of inclination. Although such an interpretation would have been consistent with the ideas of Thomas Aquinas and indeed those of Albertus Magnus, Rule IX neither specifically allowed nor disallowed this practice.[63]

The 1564 Index had deviated from the Roman Inquisition's censorship of the art of astrology, but its areas of agreement with the earlier lists on matters relating to divination and magic are of greater significance. While the Tridentine commission had attenuated some of the prohibitions contained in the Inquisition's earlier indices, they accepted a radical change in criteria used to define the boundaries of superstition. Rather than using the criteria first formulated by John XXII in the fourteenth century and subsequently elaborated by Eymerich in the *Directorium*, which stated that arts that involved making explicit contact with demons should be treated in the manner of heresy, each of the inquisitorial indices had also proscribed arts such as chiromancy that involved no ritual element. By affirming these ideas, the Tridentine Index ensured that the Inquisition's radical transformation of the criteria used to define the boundaries of orthodoxy was embedded in centralized systems of censorship. Its promulgation endorsed a new approach to the investigation of magic and divination, one based on the principles formulated by Thomas Aquinas rather than Eymerich's idiosyncratic interpretation of his thought. When considering the legitimacy of an operative art, censors could no longer simply consider an operator's actions; they were now also obliged to determine whether the effects produced by that art could be explained according to their understanding of the natural order.

The Conclusion of the Council

After the successful conclusion of Session 22, held on September 17, 1562, the end of the council had seemed to be in sight. That autumn the council instead entered a period of crisis, which began when the fathers turned to consider the decree of holy orders. The legates produced a draft decree similar to that discussed in 1552 but with the important difference that canon 7 no longer stated that a bishop was distinguished from a priest *de jure divino*. This caused outrage among some of the fathers. As John W. O'Malley has observed, "The Spaniards and the others who demanded that *jus divinum* be inserted into canon 7 were intent on enhancing episcopal authority and saw this article as crucial in attaining their goal." The legate Simonetta led the opposition to this idea, claiming that it undermined papal authority. In October 1563, the debate was given new intensity by the arrival of the long-awaited French delegation led by Cardinal de Guise. The papal party suspected that these delegates held even firmer conciliarist ideals than the Spanish, sharpening their fear that the council fathers might attempt to usurp papal authority. The session planned for November 12 in which this decree was to be discussed was delayed, but disputes over this matter continued in the General Congregation over the winter months. Divisions deepened as the fathers began once more to discuss the contentious issues of episcopal residence and reform of the College of Cardinals. Many began to fear that the divided council would be suspended again.[64]

The crisis came to a head in March 1563, when Gonzaga and Seripando died just weeks apart. Pius IV appointed Cardinal Morone, formerly Paul IV's prisoner in Castel St. Angelo, to replace Gonzaga and fulfill the role of "first president" and Cardinal Bernardo Navagero to assume Seripando's role as legate. The council had achieved little over the preceding months, but on May 12 the General Congregation recommenced its work. Morone arrived in Trent five days later, and under his stewardship the council managed to agree on a series of reforms, including the establishment of seminaries, a measure designed to enhance clerical education and improve pastoral care. Morone was also able to encourage the council to accept a compromise on the rights of bishops. The final decree stressed the importance of residence and the bishops' pastoral role. Despite the Spanish party's efforts, the decree did not acknowledge the principle that residence was required *de jure divino*. Morone next managed to secure agreement on the question of holy orders, and, in a boon for the papacy, the final text once again avoided discussion of the contested issue of the origins of episcopal authority. These measures were all passed in Session 23, held on July 15, 1563.[65]

After the successful conclusion of these sessions, Pius IV began to press Morone to bring the council to an end. The fathers still needed to discuss several outstanding doctrinal issues, including the sacrament of marriage, purgatory, and the veneration of the saints. Morone was also aware that many of the non-Italian fathers did not yet believe that the crucial matter of reform had been dealt with in sufficient detail. Under his guidance, the council moved to address these remaining issues. During Session 24, held on November 14, the fathers passed decrees on marriage and the first part of a general reform decree. As O'Malley has stressed, the reform decree's proposals amounted to a "full job description for bishops." The various provisions enjoined bishops to undertake regular visitations of their dioceses and ensure that only suitable candidates were selected for the cure of souls. Underlining the fathers' intention to place bishops at the heart of any attempt to implement the council's reforms, the decree also required archbishops to hold provincial councils every three years and bishops to hold diocesan synods annually.[66]

During this session, Morone attempted to draw the council to a conclusion by proposing to hold a final session in early December. Before it could take place, several issues still required resolution, including considering the revised index and the reform of the religious orders. By this time, news had reached Trent that Pius IV was gravely ill. Morone used the fact that the pope's death would cause the automatic suspension of the council to encourage the fathers to complete their work. The final session began on December 3 and continued into the following day. It passed doctrinal decrees concerning such issues as purgatory and the veneration of images. The fathers also acknowledged that the commission charged with revising the index had completed its work and resolved to submit the final draft to the pope for his approval prior to publication. During this final session, the fathers accepted the second part of the reform decree, which, among other matters, continued the process of defining the role of the bishop.[67]

The Tridentine fathers then proceeded to pass a further decree concerning the reform of the religious orders. As we have seen, like the question of residence, this matter raised the disputed issue of papal authority. Echoing a call frequently heard since the late fifteenth century, the decree reiterated the need for all members of the regular clergy to obey their rule. It also attempted to resolve disputes about whether regular clergy could possess property that dated back to the thirteenth century. Later chapters continued the work, begun in earlier decrees, of defining the relationship between the regular clergy and the episcopate, especially regarding the cure of souls. Contrary to the reforms of the Pauline era, these decrees reinforced the bishops' authority in relation to the mendicants and, by extension, the papacy. Chapter 11 asserted that "the individuals, whether Regulars or Seculars, who exercise that cure, shall be immediately subject, in whatsoever pertains to the said cure and the administration of the sacraments, to the jurisdiction, visitation, and correction of the bishop in whose diocese those places are situated." Chapter 12 continued this theme by stressing that the regular clergy were obliged to accept censures and interdicts passed by the ordinaries and observe any feast days they declared.[68]

The council was now finally drawing to a close. Successive legates had sought to balance the contrasting demands of the papacy, episcopate, and secular rulers, and many of the decrees were the result of awkward compromise. The papacy had managed to fend off challenges to its authority, most notably efforts to declare that the office of bishop had been established *de jure divino*. The council fathers had nevertheless passed a series of measures that not only strengthened the office of bishop, in principle placing them at the forefront of future efforts to design and implement reform, but also restrained the influence of papal agents such as the mendicants within their dioceses. It now remained only to determine how the Tridentine decrees should be applied. The penultimate decree, "On the Reception and Observance of the Council's Decrees," offered the papacy, the episcopate, and the secular rulers reason to think that they would each play the preeminent role in implementing the council's decisions, setting up a series of future power struggles. After the passage of the final decree, a brief statement formally closing the meeting and requesting papal approval of the decrees, Morone declared that the council was formally concluded.[69]

Conclusion

The events of 1549–64 directly and indirectly shaped the inquisition of learned magic in post-Tridentine Italy. During these years the Roman Inquisition became a formal part of the curia, and it established the practice of issuing Indices of Forbidden Books. The indices offered centrally determined guidance on the boundaries of orthodoxy, which shaped perceptions of superstition in the Catholic world. The inquisitorial indices of the 1550s utilized principles of censorship that were rooted in the ideas of Thomas Aquinas. While these principles were broadly consistent, they contained significant ambiguities. Three centuries of inquisitorial practice had given rise to several competing interpretations of Aquinas's writings, with the consequence that although the criteria of censorship were based on "Thomist" thought, his ideas were not always understood or interpreted consistently. Nevertheless, as a result of this development, several magical arts and practices that had previously been of contested orthodoxy were reclassified as heterodox. Inquisitors also now possessed the tools—albeit imperfect ones—that would enable them to attempt to impose their definition of orthodoxy on Italian society. These principles of censorship were broadly accepted by the authors of the Tridentine Index of 1564. Although the Tridentine fathers accepted the Roman Inquisition's criteria of censorship, they sought to circumscribe inquisitors' influence at a local level. This endeavor formed part of a broader effort to use the Council of Trent to define the respective powers of the papacy and its agents relative to those of the episcopate. Despite these efforts, conflict would break out once more between the episcopate and the Roman Inquisition in the later sixteenth century.

✷ CONCLUSION ✷

The Ambiguities of Censorship in Post-Tridentine Italy

In the early 1570s, the Jesuit Robert Bellarmine delivered a series of lectures on Thomas Aquinas's *Summa theologica* at the University of Louvain. The lectures were notable for the independent spirit in which Bellarmine interpreted Aquinas's work. When discussing predictive astrology, for example, he accepted many of Aquinas's arguments. Yet while he accepted that it was possible to predict future effects through knowledge of their causes, he appeared to harbor doubts about the doctrine of inclination.[1] When considering forms of learned magic, he once more broadly endorsed Aquinas's opinions. He accepted the possibility of manipulating natural causes to produce wondrous effects while warning of the danger posed by necromantic magic and arts such as the production of astrological talismans that involved making tacit pacts with demons. He developed his ideas in a further text bound in with lectures, *Disputatio de magia,* in which he observed that there were three types of magic: natural, demonic, and mixed. The final category referred to arts that appeared to be natural but were in fact completed by demons. There were, he continued, two recent examples of individuals discussing such mixed arts: Cornelius Agrippa's *De occulta philosophia* and Giambattista Della Porta's *Magia naturalis*. Della Porta, he continued, purported to teach about natural matters when in fact he "[taught] superstitious and dangerous magic."[2]

Bellarmine's response to Della Porta's *Magia naturalis* reflected a crucial reorientation of Christian intellectuals' understanding of learned magic, one that emphasized the danger not only of ritual activity but also of ostensibly natural arts. I suggest that this changed perspective allowed individuals such as Bellarmine to cast doubts on Della Porta's claims to produce his results through the skilled manipulation of natural phenomena and made possible his subsequent inquisitorial trial for necromancy. It is possible to locate the moment when inquisitors began to consider it necessary to censor and even prosecute the practitioners of these arts in the

1550s. Before this time, inquisitors rarely if ever investigated the practice of ostensibly natural arts but instead focused on the prosecution of individuals who explicitly invoked demons. Although Thomas Aquinas had provided a rationale for the prosecution of virtually all of these arts, his views on these matters became part of demonological discourse only during the fifteenth century and became embedded into inquisitorial practice only after the establishment of the Roman Inquisition.

My central contention in this book is that this transformation of the criteria of the censorship of magic had a radical impact on the subsequent development of science. Such a claim may still be surprising to some, but it builds on a long and rich tradition of discussing the relationship between science and magic that dates back at least as far as the researches of Lynn Thorndike. Acceptance of such material into the history of science has been a gradual process. In an essay on how to read books of secrets, William Eamon remarked on the bemusement he caused when he first attempted to discuss such material in front of historians of science in the early 1980s.[3] His anecdote acted as a foil to his main point: that the secrets literature of medieval and early modern Europe was now an accepted part of the history of science and medicine. This was a development to which he had made significant contributions. In his *Science and the Secrets of Nature*, he demonstrated that Della Porta, one of the leading professors of secrets, should be afforded a place in the history of science.[4] While he was not practicing "modern science," Della Porta was advancing philosophically grounded knowledge claims about the natural order, how it functioned, and how it could be controlled. The same was true of practitioners of ostensibly natural operative arts such as chiromancy or the production of astrological talismans. By demanding the right to adjudicate which of these knowledge claims were valid, the Roman Inquisition asserted its right to determine nature's limits. Having thus circumscribed the field in which philosophers and practitioners of the operative arts could work, they determined the acceptable boundaries of scientific knowledge.

This radical transformation of the criteria used in the censorship and prosecution of magic occurred during the religious crisis of the sixteenth century, but it was not in any meaningful sense caused by the so-called Counter-Reformation. This fact is demonstrated by the complex history of the Roman Inquisition and the Council of Trent, two institutions traditionally considered emblematic of a unified and coordinated response to the threat of Protestantism. The Roman Inquisition was founded to facilitate the pursuit of heresy until a future council could resolve the schism within Christendom. By the time the Council of Trent met, the leadership of the Roman Inquisition was pursuing a different conception of reform from that sought by a sizeable proportion of the council fathers. Supported by Emperor Charles V, many Tridentine fathers were receptive to the idea of negotiation with the Lutherans and sought to use the council to define a bishop-led program of reform driven by the renewal of pastoral ministry. The leadership of the Roman Inquisition, by contrast, opposed compromise with the Protestants and sought to eradicate whatever they perceived as heresy from within the Church. Although they too wanted to see pastoral reform, they wanted it to be directed by the papacy rather than the episcopate.

In a development separate from the Protestant crisis, the Roman Inquisition became the vehicle for the expression of the mendicants' longstanding program of reform. Operating under papal auspices, the mendicants sought to enhance pastoral care while combating heresy and superstition. After the Roman Inquisition's foundation in 1542, the Dominicans began to influence the censorship and prosecution of magic. As we have seen, the ideas that underpinned the Roman Inquisition's criteria of censorship and prosecution were developed by Thomas Aquinas in the thirteenth century but did not become important in the Dominican order's demonological discourse until the mid-fifteenth century. During the work of drafting the papal indices of the 1550s, members of the Dominican order successfully incorporated these reappropriated Thomist criteria into the heart of the Roman Inquisition's operations. Using the new universal authority of the Roman Inquisition, the Dominicans were able to advance their contested interpretations of the boundaries of superstition as orthodox Catholic belief.

The Dominicans' redefinition of the criteria of censorship occurred during the pontificate of Paul IV, a period in which inquisitorial power was expanding rapidly. As pontiff, he opposed the reconvocation of the Council of Trent; oversaw efforts to centralize reform, thereby preventing the episcopate from playing any role in determining its shape and direction; and used the Roman Inquisition to combat not only Protestants but also any perceived opponents within the Catholic Church. Pursuing this latter aim, he granted powers to the Roman Inquisition to intervene in censorship and pursue heresy at the diocesan level, powers that significantly infringed on episcopal autonomy. After Paul IV's death in August 1559, his successor, Pius IV, began to roll back some of the excesses of his reign and reconvoked the council. During the council, the Tridentine fathers reasserted episcopal authority, including the right of the council to redefine the structures of censorship and revise the Roman Inquisition's Index of Forbidden Books. The new index, promulgated in 1564, eased many of the earlier list's strictures, but significantly its compilers continued to sustain the criteria used by the Roman Inquisition to determine the boundaries of superstition.

The incorporation of the Roman Inquisition's criteria for the censorship of magic into the Tridentine Index did not, however, mark the beginning of a new unified effort to combat superstition. In the years that followed the closure of Trent, the Roman Inquisition and the episcopate continued to dispute which ideas should be condemned, contest the correct division of censorial labor, and debate who held the power to grant Catholics the right to read specific works. These arguments were directly related to and continued disputes about the respective power and authority of the papacy and the episcopate that extended back to the papal reforms of the eleventh century. They were given new potency by disputes over who was responsible for implementing Tridentine reforms and overseeing the future direction of the Church. The inquisition of magic was further complicated by efforts to determine the precise meaning of the criteria of censorship established by the 1559 and 1564 Indices. While there was a broad acceptance of the principle that in order to determine an operative art's legitimacy it was necessary to consider nature's limits, debates continued within the Roman organs of censorship about where and how to draw those limits.

Acknowledgments

During the period I have been working on this book, I have had the great pleasure to work with and learn from friends and colleagues at a number of institutions. I owe a lasting intellectual debt to Rob Iliffe, who has generously shared his erudition and helped shape my ideas in ways that I could not have imagined. At Imperial College, I benefited greatly, as both a student and a colleague, from David Edgerton's incisive advice on presenting historical arguments, both my own and those of others. During two periods teaching at the University of York, I had the great fortune to work with Simon Ditchfield, who ever since has been a consistent source of insight and encouragement. I also wish to thank former colleagues at Science, Technology and Innovation Studies at the University of Edinburgh, where I held a teaching and research fellowship. This opportunity offered me the chance to develop my research in ways that made it possible to complete this book. I am especially grateful to John Henry for reading early versions of this book's opening chapters. I would also like to thank the three reviewers who read earlier versions of this book and made comments and suggestions that have greatly improved its structure and content. Thanks too to my editor at University of Chicago Press, Randy Petilos, for guiding me through the later stages of producing this book.

While writing this book, I have also been helped in countless ways by friends and family. I would particularly like to thank Andrew Campbell, not only for all his practical help and advice but also for many hours discussing aspects of Renaissance history and engaging in scurrilous gossip. Other friends who have helped me in ways large and small over the years include Robbie Whitelaw, who kindly offered me a place to stay while I was a postdoc in Edinburgh; Jessica Price, who read the first half of this book; and Lorenza Gianfrancesco, Chiara Beccalossi, Sara Miglietti, and Justin Rivest. I would also like to give special thanks to my father-in-law, Alan Brown, who has heroically read not only the manuscript of this book but also drafts of many of my earlier publications. I am also grateful to my parents, John and Carole, and my siblings, Louise and Philip, for their support and encouragement.

Finally, I owe a tremendous debt to my own family. My children, Holly and Luke, who have endured with fortitude my frequent absences, both physical and mental, keep me grounded and richly entertained. Above all, however, I must thank my wife, Rachel, who has supported me through the highs and lows of academic life since I was a master's student. It is to her that this book is dedicated with love.

Notes

Introduction

1. Giambattista Della Porta, *Magia naturalis* (Naples, 1558), 1.1. On Della Porta's trial, see Michaela Valente, "Della Porta e l'Inquisizione: Nuovi documenti dell'archivio del Sant'Uffizio," *Bruniana e campanelliana* 5 (1999): 415–34; Neil Tarrant, "Giambattista Della Porta and the Roman Inquisition: Censorship and the Definition of Nature's Limits in Sixteenth-Century Italy," *British Journal for the History of Science* 46 (2013): 601–25. See also David J. Collins, "Learned Magic," in *The Cambridge History of Magic and Witchcraft in the West: From Antiquity to the Present*, ed. David J. Collins (Cambridge: Cambridge University Press, 2014), 332–60; Frank Klaassen, "Necromancy," in *The Routledge History of Medieval Magic*, ed. Sophie Page and Catherine Rider (London: Routledge, 2019), 201–11; Steven P. Marrone, "Magic and Natural Philosophy," in Page and Rider, *Routledge History of Medieval Magic*, 287–98.

2. Michael D. Bailey, *Fearful Spirits, Reasoned Follies: The Boundaries of Superstition in Late Medieval Europe* (Ithaca, NY: Cornell University Press, 2013), 35–70; Maijastina Kahlos, "The Early Church," in Collins, *Cambridge History of Magic*, 148–82; David J. Collins, "Scholasticism and High Medieval Opposition to Magic," in Page and Rider, *Routledge History of Medieval Magic*, 461–74.

3. Neil Tarrant, "On the Origins of Enlightenment: The Fruits of Migration in the Italian Liberal Historiographical Tradition," in *Fruits of Migration: Heterodox Italian Migrants and Central European Culture, 1550–1620*, ed. Cornel Zwierlein and Vincenzo Lavenia (Brill: Leiden, 2018), 362–81; Neil Tarrant, "Science, Religion and Italy's Seventeenth-Century Decline: From Francesco De Sanctis to Benedetto Croce," *Zygon* 54 (2019): 1125–44.

4. Tarrant, "Origins of Enlightenment," 367–78; "Science, Religion," 1130–31, 1136–41. For an indication of the wealth of secondary literature on the Italian Reformation, see John Tedeschi, *The Italian Reformation of the Sixteenth Century and the Diffusion of Renaissance Culture: A Bibliography of the Secondary Literature (ca. 1750–1997)*, in association with J. M. Lattis (Modena: F.C. Panini, 2000), and especially Massimo Firpo's "Historiographical Introduction," xviii–xlix. See also Silvana Seidel Menchi, "Italy," in *The Reformation in National Context*, ed. Bob Scribner, Roy Porter, and Mikuláš Teich (Cambridge: Cambridge University Press, 1994), 181–201; Adriano Prosperi, "Italy," in *The Oxford Encyclopedia of the Renaissance*, vol. 2, ed. Hans J. Hillebrand (Oxford: Oxford University Press, 1996), 324–29.

5. Neil Tarrant, "Censoring Science in Sixteenth-Century Italy: Recent (and Not-

So-Recent) Research," *History of Science* 52 (2014): 1–27; Tarrant, "Science, Religion," 1133–41.

6. For a useful, albeit now somewhat dated overview, see Paula Findlen, "Science and Society," in *Early Modern Italy*, ed. John A. Marino (Oxford: Oxford University Press, 2002), 166–87.

7. Ugo Baldini and Leen Spruit, eds., *Catholic Church and Modern Science: Documents from the Archives of the Holy Office and the Index*, vol. 1, *Sixteenth-Century Documents* (Rome: Libreria Editrice Vaticana: 2009), 70–71.

8. For further discussion of this issue, see Tarrant, "Censoring Science," 16–22.

9. Christine Caldwell Ames, *Righteous Persecution: Inquisition, Dominicans and Christianity in the Middle-Ages* (Philadelphia: University of Pennsylvania Press, 2009).

10. For a discussion of the categories applied to this period, see John W. O'Malley, *Trent and All That: Renaming Catholicism in the Early Modern Era* (Cambridge, MA: Harvard University Press, 2000), and for a concise summary of Jedin's work and influence, see 46–71.

11. For an overview of confession building or confessionalization, see Ute Lotz-Heumann, "The Concept of 'Confessionalisation': A Historiographical Paradigm in Dispute," *Memoria y Civilización* 4 (2001): 93–114.

12. Simon Ditchfield, "Tridentine Catholicism," in *The Ashgate Companion to the Counter-Reformation*, ed. Alexandra Bamji, Geert H. Janseen, and Mary Laven (Farnham: Ashgate, 2013), 15–18. John W. O'Malley, *Trent: What Happened at the Council* (Harvard University Press: Cambridge MA, 2013), 1–22 and 248–75. On the Second Vatican Council, see John W. O'Malley, *What Happened at Vatican II* (Cambridge, MA: Belknap, 2008).

13. Massimo Firpo, *Inquisizione Romana e Controriforma: Studi sul Cardinale Giovanni Morone (1509–1580) e il suo processo di eresia*, 2nd ed. (Brescia: La Morcelliana, 2005); Massimo Firpo, *La presa di potere dell'Inquisizione Romana, 1550–1553* (Rome: Laterza, 2014).

14. Gigliola Fragnito, *La bibbia al rogo: La censura ecclesiastica e i volgarizzamente della Scrittura (1471–1605)* (Bologna: Il Mulino, 1997); Gigliola Fragnito, *Proibito capire: La chiesa e il volgare nella prima età moderna* (Bologna: Il Mulino, 2005); Gigliola Fragnito, *Rinascimento perduto: La letteratura italiana sotto gli occhi dei censori (secoli XV–XVII)* (Bologna: Il Mulino, 2019); Vittorio Frajese, *Nascita dell'Indice: La censura ecclesiatica dal Rinascimento alla Controriforma*, 2nd ed. (Brescia: Morcelliana, 2008).

15. Adriano Prosperi, *Tribunali della coscienza: Inquisitori, confessori, missionari*, new ed. (Einaudi: Turin, 1996).

16. Edward Peters, *Inquisition* (Berkeley: University of California Press, 1989); Andrea Del Col, *L'Inquisizione in Italia, dal XII al XXI secolo* (Milan: Mondadori, 2006); Elena Brambilla, *Alle origini del Sant' Uffizio: Penitenzia confessione, e giustizia spiritual dal medioevo al XVI secolo* (Bologna: Il Mulino, 2000).

17. Simon Ditchfield, "Of Dancing Cardinals and Mestizo Madonnas: Reconfiguring the History of Roman Catholicism in the Early Modern Period," *Journal of Early Modern History* 8 (2004): 408.

18. Rosalind B. Brooke, *The Coming of the Friars* (London: George Allen and Unwin, 1975); William A. Hinnebusch, *The History of the Dominican Order*, 2 vols. (New York: Alba House, 1966–1973); Augustine Thomson, *Revival Preachers and Politics in Thirteenth-Century Italy* (Oxford: Clarendon, 1992); C. H. Lawrence, *The Friars: The Impact of the Early Mendicant Movement on Western Society* (London: Longman,

1994); Michael Robson, *The Franciscans in the Middle Ages* (Woodbridge: Boydell, 2006); Frances Andrews, *The Other Friars: The Carmelite, Augustinian and Pied Friars in the Middle Ages* (Woodbridge: Boydell, 2006); Donald S. Prudlo, ed., *The Origin, Development, and Refinement of Medieval Religious Mendicancies* (Leiden: Brill, 2011).

19. Colin Morris, *Papal Monarchy: The Western Church from 1050–1250* (Oxford: Oxford University Press, 1989).

20. For an overview, see Ronald J. Stansbury, "Preaching and Pastoral Care in the Middle Ages," in *A Companion to Pastoral Care in the Middle Ages (1200–1500)*, ed. Ronald J. Stansbury (Leiden: Brill, 2010), 23–39. On preaching, see Katherine Jansen, "The Word and Its Diffusion," in *The Cambridge History of Christianity*, vol. 4, *Christianity in Western Europe, c.1100–c.1500*, ed. Miri Rubin and Walter Simons (Cambridge: Cambridge University Press, 2009), 114–32; as it relates to magic, see Kathleen Kamerick, "Pastoral Literature and Preaching," in Page and Rider, *Routledge History of Medieval Magic*, 475–86.

21. Roberto Rusconi, *L'ordine dei peccati: La confessione tra Medieoevo ed età moderna* (Bologna: Il Mulino, 2002).

22. Ute Lotz-Heumann, "Imposing Church and Social Discipline," in *The Cambridge History of Christianity*, vol. 6, *Reform and Expansion, 1500–1600*, ed. R. Po-chia Hsia (Cambridge: Cambridge University Press, 2007), 249–50.

23. Andrew Traver, "The Forging of an Intellectual Defense of Mendicancy in the Medieval University," in Prudlo, *Origin, Development, and Refinement*, 157–95.

24. Francis Oakley, *The Conciliarist Tradition: Constitutionalism in the Catholic Church, 1300–1870* (Oxford: Oxford University Press, 2003); Paul Valliere, *Conciliarism: A History of Decision-Making in the Church* (Cambridge: Cambridge University Press, 2012).

25. Luigi Firpo, "Filosofia italiana e controriforma," *Rivista di filosofia* 41 (1950): 152–53.

26. Fragnito, *Proibito capire*, 9; Elena Bonora, "The Takeover of the Roman Inquisition," in *A Companion to Heresy Inquisitions*, ed. Donald S. Prudlo (Leiden: Brill, 2019), 267–76.

27. Guido Ruggiero, *Binding Passions: Tales of Magic, Marriage and Power at the End of the Renaissance* (New York: Oxford University Press, 1993), 9–10; Menchi, "Italy," 194; Anne Jacobson Schutte, *Aspiring Saints: Pretense of Holiness, Inquisition, and Gender in the Republic of Venice, 1618–1750* (Baltimore: Johns Hopkins Press, 2001), 28–29; Anne Jacobson Schutte, "Religion, Spirituality, and the Post-Tridentine Church," in Marino, *Early Modern Italy*, 127; Christopher Black, *The Italian Inquisition* (New Haven, CT: Yale University Press, 2009), 231–32.

28. Matteo Duni, *Under the Devil's Spell: Witches, Sorcerers, and the Inquisition in Renaissance Italy* (Syracuse, NY: Syracuse University Press), 32–34, quote 33.

29. On the materials available to scholars, see John Tedeschi, "The Dispersed Archives of the Roman Inquisition," in *The Inquisition in Early Modern Europe: Studies in Sources and Methods*, ed. Gustav Henningsen and John Tedeschi (DeKalb: Northern Illinois University Press, 1986), 13–32.

30. William Monter and John Tedeschi, "Towards a Statistical Profile of the Italian Inquisitions: Sixteenth to Eighteenth Centuries," in Henningsen and Tedeschi, *Inquisition in Early Modern Europe*, 130–57.

31. Frajese, *Nascita dell'Indice*, 73–80.

32. See, for example, Jeremy Cohen, *The Friars and the Jews: The Evolution of Medi-*

eval Anti-Judaism (Ithaca, NY: Cornell University Press, 1982); Bernadette Paton, *Preaching Friars and the Civic Ethos: Sienna, 1380–1480* (London: Centre for Medieval Studies, Queen Mary and Westfield College, University of London, 1992); Franco Mormando, *The Preacher's Demons: Bernardino of Siena and the Social Underworld of Early Renaissance Italy* (Chicago: University of Chicago Press, 1999); Nirit Ben-Aryeh Debby, *Renaissance Florence in the Rhetoric of Two Popular Preachers: Giovanni Dominici (1356–1419) and Bernardino da Siena (1380–1444)* (Turnhout: Brepols, 2001); Steven McMichael and Susan Myers, eds., *Friars and Jews in the Middle Ages and Renaissance* (Leiden: Brill, 2004); Robin Vose, *Dominicans, Muslims and Jews in the Medieval Kingdom of Aragon* (Cambridge: Cambridge University Press, 2009).

33. Important studies include Norman Cohn, *Europe's Inner Demons: An Enquiry Inspired by the Great Witch-Hunt* (London: Heinemann for Sussex University Press, 1975); Richard Kieckhefer, *The European Witch Trials: Their Foundations in Popular and Learned Culture* (London: Routledge, 1976); Edward Peters, *The Magician, the Witch, and the Law* (Hassocks: University of Pennsylvania Press, 1978).

34. H. C. Erik Midelfort, "Witch Craze: Beyond the Legend of the Panic," *Magic, Ritual, and Witchcraft* 6 (2011): 11–33, esp. 16.

35. Neil Tarrant, "Between Aquinas and Eymerich: The Roman Inquisition's Use of Dominican Thought in the Censorship of Alchemy," *Ambix* 65 (2018): 210–31; Neil Tarrant, "The Ambiguities of Censorship in Post-Tridentine Italy: The Case of Physiognomy," *Preternature* 8 (2019): 171–201.

36. Kieckhefer, *European Witch Trials*, 20–22. On the inquisition of diabolical sects of witches in late fifteenth- and early sixteenth-century Italy, see Del Col, *L'Inquisizione*, 176–217; Michael Tavuzzi, *Renaissance Inquisitors: Dominican Inquisitors and Inquisitorial Districts in Northern Italy, 1474–1527* (Leiden: Brill, 2007), 149–208; Duni, *Under the Devil's Spell*, 27–38. On the persistence of the elaborated stereotype of the witch in Italy, see Carlo Ginzburg, *The Night Battles: Witchcraft and Agrarian Cults in the Sixteenth and Seventeenth Centuries*, trans. John Tedeschi and Anne C. Tedeschi (London: Routledge, 1983); Prosperi, *Tribunali della coscienza*, 368–99; C. Black, *Italian Inquisition*, 231–54.

37. John Tedeschi, "Inquisitorial Law and the Witch," in *Early Modern European Witchcraft: Centres and Peripheries*, ed. Bengt Ankaloo and Gustav Henningsen (Oxford: Oxford University Press, 1990), 83–118. The idea that the Roman Inquisition ended the Italian witch hunts is now well established. See, for example, C. Black, *Italian Inquisition*, 231–34; Tamar Herzig, "Bridging North and South: Inquisitorial Networks and Witchcraft Theory on the Eve of the Reformation," *Journal of Early Modern History* 12 (2008): 361–62; Midelfort, "Witch Craze," 14–16.

38. Ruth Martin, *Witchcraft and the Inquisition in Venice, 1550–1650* (Oxford: Basil Blackwell, 1989), Chapter 2, quote on p. 55. See also, however, the remarks of Jonathan Seitz, *Witchcraft and Inquisition in Early Modern Venice* (Cambridge: Cambridge University Press, 2011), 197–201, who downplays Eymerich's importance to the investigation of magic and witchcraft.

39. Francesco Beretta, "Orthodoxie philosophique et Inquisition Romaine aux 16ᵉ–17ᵉ siècles: Un essai d'interpretation," *Historia philosophica* 3 (2005): 67–96.

40. For a critique of Weber's historiographical influence, see Alexandra Walsham, "The Reformation and the 'Disenchantment of the World' Reassessed," *Historical Journal* 51 (2008): 497–528.

41. John Henry, "The Fragmentation of Renaissance Occultism and the Decline of

Magic," *History of Science* 46 (2008): 1–48; for his comments on the various churches' influence, see 16.

42. Stuart Clark, *Thinking with Demons: The Idea of Witchcraft in Early Modern Europe* (Oxford: Oxford University Press, 1997), esp. pt. 2, "Science"; Robert Bartlett, *The Natural and the Supernatural in the Middle Ages* (Cambridge: Cambridge University Press, 2008); William R. Newman, "Technology and Alchemical Debate in the Late Middle Ages," *Isis* 80 (1989): 423–445; William R. Newman, "Art, Nature and Demons: The Case of the *Malleus Maleficarum* and Its Medieval Sources," in *The Artificial and the Natural: An Evolving Polarity*, ed. Bernadette Bensaude-Vincent and William R. Newman (Cambridge, MA: MIT Press, 2007), 109–34.

43. I have drawn these categories from Clark, *Thinking with Demons*, 151–60.

44. Clark, *Thinking with Demons*, 158–60, 161–78.

45. For a concise overview of modern sociologically informed history of science, see Jan Golinski, *Making Natural Knowledge: Constructivism and the History of Science* (Cambridge: Cambridge University Press, 1998); see also Steven Shapin, *Never Pure: Historical Studies of Science as If It Was Produced by People with Bodies, Situated in Time, Space, Culture and Society, and Struggling for Credibility and Authority* (Baltimore: Johns Hopkins University Press, 2010).

46. Historians continue to dispute the causes of these developments, considering whether they reflected Christian responses to real changes—for example, the rise of heresy as a response to a yearning for reform—or whether they reflected a change in the perceptions of the dominant Catholic majority. For a discussion of heresy that emphasizes the proliferation of heretics, see Malcolm Lambert, *Medieval Heresy: Popular Movements from the Gregorian Reform to the Reformation*, 3rd ed. (Oxford: Blackwell, 2002). A pivotal work of revision was offered by R. I. Moore, *The Formation of a Persecuting Society: Authority and Deviance in Western Europe, 950–1250*, paperback ed. (Malden, MA: Blackwell, 1990). For critiques of Moore's work, see, for example, John Christian Laursen and Cary J. Nederman, eds., *Beyond the Persecuting Society: Religious Toleration before the Enlightenment* (Philadelphia: University of Pennsylvania Press, 1998); Christine Caldwell Ames, "Does Inquisition Belong to Religious History?," *American Historical Review* 110 (2005): 11–37.

47. On Observant reform, see James D. Mixson, "Religious Life and Observant Reform in the Fifteenth Century," *History Compass* 11 (2013): 201–14; James D. Mixson and Bert Roest, eds., *A Companion to Observant Reform in the Late Middle Ages and Beyond* (Leiden: Brill, 2015).

Chapter One

1. "*Quod super nonnullis*," in *Sexti decretalium D. Bonifacii Papae VIII* (Lyons, 1584), 5.2.8.4, col. 622.

2. On science in early medieval Western Europe, see David Lindberg, *The Beginnings of Western Science: The European Scientific Tradition in Philosophical, Religious and Institutional Context, 600 B.C. to A.D. 1450* (Chicago: University of Chicago Press, 1992), 151–59, 183–213; Edward Grant, *A History of Natural Philosophy: From the Ancient World to the Nineteenth Century* (Cambridge: Cambridge University Press, 2007), 95–142; Stephen McCluskey, "Natural Knowledge in the Early Middle Ages," in *The Cambridge History of Science*, vol. 2, *Medieval Science*, ed. David C. Lindberg and Michael H. Shank (Cambridge: Cambridge University Press, 2013), 286–301.

3. For a survey of changing historiographical conceptions of the tenth century, see John Howe, "Re-forging the 'Age of Iron,' Part 1: The Tenth Century as the End of the Ancient World?," *History Compass* 8 (2010): 866–87; John Howe, "Re-forging the Age of Iron, Part 2: The Tenth Century in a New Age?," *History Compass* 8 (2010): 1000–1022.

4. On the development of schools and the universities, see Hilde de Ridder-Symoens, ed., *A History of the University in Europe*, vol. 1, *Universities in the Middle Ages* (Cambridge: Cambridge University Press, 1992); Olaf Pederson, *The First Universities:* Studium Generale *and the Origins of University Education in Europe* (Cambridge: Cambridge University Press, 1997). On the foundations and distinctive characteristics of the early Italian universities, see Paul F. Grendler, *The Universities of the Italian Renaissance* (Baltimore: Johns Hopkins University Press, 2002), 3–40.

5. For a discussion of the intellectual history of the Latin West, see Michael Haren, *The Western Intellectual Tradition from Antiquity to the Thirteenth Century* (Basingstoke: Macmillan, 1992); Marcia L. Colish, *Medieval Foundations of the Western Intellectual Tradition, 400–1400* (New Haven, CT: Yale University Press, 1997).

6. On the translation movement and its impact on science, see Lindberg, *Beginnings*, 151–59, 183–213; Grant, *History of Natural Philosophy*, 130–78; Charles Burnett, "The Twelfth-Century Renaissance," in Lindberg and Shank, *Cambridge History of Science*, 2:365–84.

7. Lindberg, *Beginnings*, 151–59.

8. On the translation movement, see Lindberg, *Beginnings*, 203–13; Charles Burnett, "Translation and Transmission of Greek and Arabic Science to Latin Christendom," in Lindberg and Shank, *Cambridge History of Science*. 2:341–64. See too Charles H. Lohr, "The Medieval Interpretation of Aristotle," in *The Cambridge History of Later Medieval Philosophy: From the Rediscovery of Aristotle to the Disintegration of Scholasticism, 1100–1600*, ed. Norman Kretzman, Anthony Kenny, and Jan Pinborg (Cambridge: Cambridge University Press, 1982), 80–98.

9. For an overview of medieval medicine, see, for example, Nancy G. Siraisi, *Medieval and Early Renaissance Medicine: An Introduction to Knowledge and Practice* (Chicago: University of Chicago Press, 1990); Vivian Nutton, "Medicine in Medieval Western Europe, 1000–1500," in *The Western Medical Tradition: 800 BC to AD 1800*, ed. Lawrence I. Conrad, Michael Neve, Vivian Nutton, Roy Porter, and Andrew Wear (Cambridge: Cambridge University Press, 1995), 139–206.

10. On magic in the early medieval period, see Valerie I. E. Flint, *The Rise of Magic in Medieval Europe* (Princeton, NJ: Princeton University Press, 1991); Yitzhak Hen, "The Early Medieval West," in Collins, *Cambridge History of Magic*, 183–206. On the practice of popular magic in medieval Europe, see Richard Kieckhefer, *Magic in the Middle Ages*, 2nd ed. (Cambridge: Cambridge University Press, 2014), 56–94; Catherine Rider, "Common Magic," in Collins, *Cambridge History of Magic*, 303–31.

11. For an overview of the development of learned magic in this period, see Jean-Patrice Boudet, *Entre science et* Nigromance: *Astrologie, divination et magie dans l'Occident medievale (xii*[e]*–vx*[e] *siècle)* (Paris: Publications de la Sorbonne, 2006), 33–155.

12. On the development of astrology and astrological images, see Boudet, *Entre science et* Nigromance, 44–82; Nicolas Weill-Parot, *Les "images astrologiques" au Moyen Âge et à la Renaissance: Spéculations intellectuelles et pratiques magiques (XIIe–XVe siècle)* (Paris: Champion, 2002); John North, "Astronomy and Astrology," in Lindberg and Shank, *Cambridge History of Science*, 2:473–74; Charles Burnett, *Magic and Divi-*

nation in the Middle Ages: Texts and Techniques in the Islamic and Christian Worlds (Aldershot: Ashgate, 1996); Frank Klaassen, *The Transformations of Magic: Illicit Learned Magic in the Later Middle Ages and Renaissance* (University Park: Pennsylvania State University Press, 2013); H. Darrel Rutkin, *Sapientia Astrologica: Astrology, Magic and Natural Knowledge, ca. 1250–1800*, vol. 1, *Medieval Structures (1250–1500): Conceptual, Institutional, Socio-political, Theologico-religious, Cultural* (Cham: Springer, 2019).

13. On physiognomy, see Martin Porter, *Windows of the Soul: The Art of Physiognomy in European Culture, 1470–1780* (Oxford: Oxford University Press, 2005); Charles Burnett, "The Earliest Chiromancy in the West," in *Magic and Divination*, essay 10, 189–95.

14. On alchemy, see Lawrence M. Principe, *The Secrets of Alchemy* (Chicago: University of Chicago Press, 2013), chaps. 1–3; William R. Newman, "Medieval Alchemy," in Lindberg and Shank, *Cambridge History of Science*, 2:385–403.

15. Kieckhefer, *Magic in the Middle Ages*, 151–75. For a more detailed exploration of clerical magic, see, for example, Sophie Page, *Magic in the Cloister: Pious Motives, Illicit Interests, and Occult Approaches to the Medieval Universe* (University Park: Pennsylvania State University Press, 2013).

16. Steven Clucas, "Exorcism, Conjuration and the Historiography of Early Modern Ritual Magic," in *Aisthetics of the Spirits: Spirits in Early Modern Science, Religion, Literature and Music*, ed. Steffen Schneider (Göttingen: Vandenhoeck and Ruprecht, V&R Unipress, 2015), 261–85. See also Claire Fanger, ed., *Conjuring Spirits: Texts and Traditions of Medieval Ritual Magic* (University Park: Pennsylvania State University Press, 1998). On the relationship between ancient and medieval forms of theurgy, see Claire Fanger, "Introduction: Theurgy, Magic, Mysticism," in *Invoking Angels: Theurgic Ideas and Practices, Thirteenth to Seventeenth Centuries*, ed. Claire Fanger (University Park: Pennsylvania State University Press, 2012), 15; Boudet, *Entre science et* Nigromance, 137–55.

17. Cohn, *Europe's Inner Demons*, 164–73; Kieckhefer, *Magic in the Middle Ages*, 151–75.

18. Kieckhefer, *Magic in the Middle Ages*, 177–80; Kahlos, "Early Church," 148–82.

19. Harold J. Berman, *Law and Revolution: The Formation of the Western Legal Tradition* (Cambridge, MA: Harvard University Press, 1983), 199–224. On the church fathers and superstition, see Boudet, *Entre science et* Nigromance, 205–10; Bailey, *Fearful Spirits*, 38–54; Tim Hegedus, *Early Christianity and Astrology* (New York: Peter Lang, 2007).

20. Burchard of Worms, *Decretorum Libri XX*, 10.1 and 10.6, in *Burchardi Vormatiensis Episcopi Opera Omnia*, ed. J.-P. Migne (Paris: Garnier Fratres, 1880).

21. On the misattribution of this passage to Jerome, see Henry Ansgar Kelly, "Canon Law and Chaucer on Licit and Illicit Magic," in *Law and the Illicit in Medieval Europe*, ed. Ruth Mazo Karras, Joel Kaye, and E. Ann Matter (Philadelphia: University of Pennsylvania Press, 2008), 12–13.

22. Burchard of Worms, *Decretorum*, 10.42–44; c.f. Barney, Steven A., W. J. Lewis, J. A. Beach, and Oliver Berghof, trans., *The Etymologies of Isidore of Seville* (Cambridge: Cambridge University Press, 2006), 8.9.4–35; on Kalends, see 10.15–17.

23. Ivo of Chartres, *Decretum*, 11.13, citing *De doctrina christiana*, 2.19–23; and 11.67–69, citing Isidore of Seville, *Etymologiae*, 8.9.4–35 as Augustine, in *Sancti Ivonis Cartonensis episcopi opera omnia*, ed. J.-P. Migne (Paris: J.-P. Migne, 1889).

24. Ernst-Dieter Hehl, "War, Peace and the Christian Order," in *The New Cam-*

bridge Medieval History, vol. 4, pt. 1, *c.1024–c.1198*, ed. David Luscombe and Jonathan Riley-Smith (Cambridge: Cambridge University Press, 2004), 185–228.

25. Rosamund McKitterick, "The Church," in *The New Cambridge Medieval History*, vol. 3, *c.900–c.1024*, ed. Timothy Reuter (Cambridge: Cambridge University Press, 2000), 130–36; I. S. Robinson, "Reform and the Church, 1073–1122," in Luscombe and Riley-Smith, *New Cambridge Medieval History*, 4.1:272–76; Uta-Renate Blumenthal, "The Papacy, 1024–1122," in Luscombe and Riley-Smith, *New Cambridge Medieval History*, 4.2:11–12.

26. Claudio Azzara, "The Papacy," in *Italy in the Early Middle Ages*, ed. Christina La Rocca (Oxford: Oxford University Press, 2002), 109–15; Thomas F. X. Noble, "The Papacy in the Eighth and Ninth Centuries," in *The New Cambridge Medieval History*, vol. 2, *c.700–c.900*, ed. Rosamond McKitterick (Cambridge: Cambridge University Press, 1995), 563–86.

27. Kathleen G. Cushing, *Reform and the Papacy in the Eleventh Century: Spirituality and Social Change* (Manchester: Manchester University Press, 2005); I. Robinson, "Reform and the Church," 268–76. On the transformation of the eleventh-century papacy, see Blumenthal, "Papacy," 14–16. On the development of the Papal States, see Brenda Bolton, "Papal Italy," in *Italy in the Central Middle Ages, 1000–1300*, ed. David Abulafia (Oxford: Oxford University Press, 2004), 82–85. For a detailed study of the roots of the eleventh-century reform movement, see John Howe, *Before the Gregorian Reform: The Latin Church at the Turn of the First Millennium* (Ithaca, NY: Cornell University Press, 2016).

28. Berman, *Law and Revolution*, 93–99; I. Robinson, "Reform and the Church," 268–86; Morris, *Papal Monarchy*, 32–33; H. E. J. Cowdrey, "The Structure of the Church, 1024–1073," in Luscombe and Riley-Smith, *New Cambridge Medieval History*, 4.1:229–67.

29. For an overview of the history of councils, see Valliere, *Conciliarism*, 49–118. See also Blumenthal, "Papacy," 17–32; Berman, *Law and Revolution*, 85–119.

30. Berman, *Law and Revolution*, 143–51; Grendler, *Universities of the Italian Renaissance*, 431–32.

31. Berman, *Law and Revolution*, 199–204.

32. I. Robinson, "Reform and the Church," 274.

33. Bernard Hamilton, "Religion and the Laity," in Luscombe and Riley-Smith, *New Medieval Cambridge History*, 4.1:499–500, 508–10.

34. For the Lateran decrees, see Norman P. Tanner, ed., *Decrees of the Ecumenical Councils*, vol. 1, *Nicaea I to Lateran V* (London: Sheed and Ward, 1990), 190–203.

35. Hamilton, "Religion and the Laity," 499–500, 508–10.

36. Canon 23 of Lateran II, which condemned as heretics those who denied the Eucharist, infant baptism, priesthood, and legitimate marriage, was a rare example of a doctrinal decree. Canon 13 of Lateran I condemned the production of counterfeit metals as fraudulent, but the fathers made no reference to superstitious or magical arts. Canons of Lateran I and II in Tanner, *Decrees*, 1:190–203.

37. Gratian, *Decretum gratiani* (Lyons, 1589), 2.26.1–5. For further discussion of Gratian and magic, see Peters, *Magician, Witch, and Law*, 71–72; Berman, *Law and Revolution*, 143–51; Grendler, *Universities of the Italian Renaissance*, 431–32; Boudet, *Entre science et* Nigromance, 241–44; Bailey, *Fearful Spirits*, 57–58.

38. Peters, *Inquisition*, 44–47; Michael Frassetto, "Precursors to Religious Inqui-

sitions: Anti-heretical Efforts to 1184," in Prudlo, *Companion to Heresy Inquisitions*, 41–72.

39. For an overview, see Lucy J. Sackville, "The Church's Institutional Response to Heresy in the 13th Century," in Prudlo, *Companion to Heresy Inquisitions*, 108–40. For a more detailed discussion of transformations in Christian thought about the Jews, see Gavin Langmuir, *Towards a Definition of Antisemitism* (Berkeley: University of California Press, 1990).

40. Kenneth R. Stow, "The Church and the Jews," in *The New Cambridge Medieval History*, vol. 5, *c.1198–c.1300*, ed. David Abulafia (Cambridge: Cambridge University Press, 1999), 204–10.

41. On Waldo and his followers, see Lambert, *Medieval Heresy*, 70–96; Gabriel Audisio, *The Waldensian Dissent: Persecution and Survival, c.1170–c.1570* (Cambridge: Cambridge University Press, 1999), 6–25; Euan Cameron, *Waldenses: Rejections of Holy Church in Medieval Europe* (Oxford: Blackwell, 2000), 11–35.

42. Historians debate the existence of the Cathars. Some argue that the Cathars constituted a coherent heretical sect. See Hamilton, "Religion and the Laity," 503–4; Lambert, *Medieval Heresy*, 60–66, 108–50. Others reject this view. See Mark Gregory Pegg, *A Most Holy War: The Albigensian Crusade and the Battle for Christendom* (Oxford: Oxford University Press, 2008); R. I. Moore, *The War on Heresy: Faith and Power in Medieval Europe* (London: Profile Books, 2012).

43. Lateran III, Canon 27, in Tanner, *Decrees*, 1:224.

44. Lateran III, Canon 27, in Tanner, *Decrees*, 1:225.

45. Peters, *Inquisition*, 47–48.

46. Lateran IV, Canons 1, 13, and 21, in Tanner, *Decrees*, 1:230, 242, 245; Rusconi, *L'ordine dei peccati*, 29–41.

47. Lateran IV, Canons 67–70, in Tanner, *Decrees*, 1:265–67; Stow, "Church and the Jews," 207–8.

48. Lateran IV, Canon 3, in Tanner, *Decrees*, 1:233–35; Bernard Hamilton, "The Albigensian Crusade and Heresy," in Abulafia, *New Cambridge Medieval History*, 5:164–81.

49. Brooke, *Coming of the Friars*, 20–39, 89–113; Hinnebusch, *History of the Dominican Order*, 1:13–38; Lawrence, *Friars*, 26–42, 65–85; Robson, *Franciscans*, 10–21.

50. Cohen, *Friars and the Jews*, esp. chaps. 1–4; Stow, "Church and the Jews," 210–12, Vose, *Dominicans, Muslims and Jews*, esp. chaps. 4–5.

51. Rusconi, *L'ordine dei peccati*, 105–60; Robson, *Franciscans*, 48–57, 70–71.

52. R. N. Swanson, "The 'Mendicant Problem' in the Later Middle Ages," in *The Medieval Church: Universities, Heresy and the Religious Life, Essays in Honour of Gordon Leff*, ed. Pete Biller and Barrie Dobson (Woodbridge: Boydell, 1999), 217.

53. On the friars' preaching, see Lawrence, *Friars*, 19–23; on their use of education, see Roger French and Andrew Cunningham, *Before Science: The Invention of the Friars' Natural Philosophy* (Aldershot: Ashgate, 1996), 99–172.

54. On the Dominican educational system, see Hinnebusch, *History of the Dominican Order*, 2:19–98; Lawrence, *Friars*, 70–75, 127–34; French and Cunningham, *Before Science*, 150–60; M. Michèle Mulchahey, *"First the Bow Is Bent in Study . . .": Dominican Education before 1350* (Toronto: Pontifical Institute of Medieval Studies, 1998), 3–71.

55. On the Franciscan *studium*, see J. C. Murphy, "The Early Franciscan Studium at

the University of Paris," in *Studium generale: Studies Offered to Astrik L. Gabriel*, ed. Leslie S. Domonkos and Robert J. Schneider (Notre Dame, IN: Medieval Institute, University of Notre Dame, 1967), 159–203; Robson, *Franciscans*, 58–65.

56. Traver, "Forging of an Intellectual Defense," 158–60; French and Cunningham, *Before Science*, 160; Monika Asztalos, "The Faculty of Theology," in de Ridder-Symoens, *History of the University*, 1:415.

57. Murphy, "Early Franciscan Studium," 190–97; Lawrence, *Friars*, 152–59; Traver, "Forging of an Intellectual Defense," 157–95; Andrew Traver, "Secular and Mendicant Masters of the Faculty of Theology at the University of Paris, 1505–1523," *Sixteenth Century Journal* 26 (1995): 141.

58. Lyon II, Canon 23, in Tanner, *Decrees*, 1:326–27.

59. James B. Given, *Inquisition and Medieval Society: Power, Discipline, and Resistance in Languedoc* (Ithaca, NY: Cornell University Press, 1997), 13–15.

60. On Dominicans as inquisitors, see Wolfram Hoyer, ed., *Praedicatores, Inquisitores I: The Dominicans and the Medieval Inquisition—Acts of the First International Seminar on the Dominicans and the Inquisition, Rome 23–25 February 2002* (Rome: Instituto Storico Domenico, 2004).

61. Caldwell Ames, *Righteous Persecution*.

62. Peters, *Inquisition*, 57–58, 68. For a discussion of the existence of a "Medieval Inquisition," see Richard Kieckhefer, "The Office of the Inquisition and Medieval Heresy: The Transition from Personal to Institutional Jurisdiction," *Journal of Ecclesiastical History* 46 (1995): 36–61. See also John Arnold, *Inquisition and Power: Catharism and the Confessing Subject in Medieval Languedoc* (Philadelphia: University of Pennsylvania Press, 2001), 77–79; Given, *Inquisition and Medieval Society*, 15. On the organization of inquisition in Italy, see Jill Moore, *Inquisition and Its Organisation in Italy, 1250–1350* (Woodbridge: York Medieval, 2019), who offers valuable qualifications to Kieckhefer's thesis by showing that individual tribunals enjoyed significant degrees of institutionalization.

63. Lucy J. Sackville, "The Inquisitor's Manual at Work," *Viator* 44 (2013): 201–5. For further discussion of the genre of the manual, see Given, *Inquisition and Medieval Society*, 44–51.

64. Peters, *Inquisition*, 52–58. For the *Ordo processus narbonensis*, see "A Manual for Inquisitors," in Walter L. Wakefield, *Heresy, Crusade and Inquisition in Southern France, 1100–1250* (London: George Allen and Unwin, 1974), 250–58. For a discussion of the authorship of this text, see Sackville, "Inquisitor's Manual," 207n21. For the text of the *Summa de catharis et pauperibus de Lugduno*, see "The *Summa* of Ranierius Sacconi," in *Heresies of the High Middle Ages: Selected Sources Translated and Annotated*, ed. Walter W. Wakefield and Austin P. Evans (New York: Columbia University Press, 1969), 329–46.

65. "Pestis inquisitores haereticae, a sede apostolica deputati, de divinationibus, aut sortilegiis, nisi haeresim saperent manisfeste, intromittere se non debent, nec punire talia exercentes, sed eos relinquere suis iudicibus puniendos." *Sexti decretalium D. Bonifacii Papae VIII* (Lyons, 1584), 5.2.8.4.

Chapter Two

1. French and Cunningham, *Before Science*.

2. Lindberg, *Beginnings*, 216–28.

3. Augustine, *De civitate Dei*, 21.6, in *Sancti Aurelii Augustini, Hipponensis Episcopi, Opera omnia*, ed. J.-P. Migne, vol. 7 (Paris: Garnier Fratres, 1841).

4. French and Cunningham, *Before Science*, 127–42.

5. Lindberg, *Beginnings*, 216–18; Theodore Crowley, *Roger Bacon: The Problem of the Soul in His Philosophical Commentaries* (Louvain: Editions de l'Institut Supérior de Philosophie, 1950), 23–26, 76–78.

6. French and Cunningham, *Before Science*, 173–83; Hinnebusch, *History of the Dominican Order*, 2:19–44.

7. For a brief biography, see William A. Wallace, "Albertus Magnus, Saint," in *Dictionary of Scientific Biography*, vol. 1, ed. Charles Coulston Gillespie (New York: Charles Scribner's Sons, 1970), 99–103. On his role in developing the Dominican philosophical curriculum, see Mulchahey, *"First the Bow Is Bent,"* 252–67. For further discussion of Albertus's scientific work, see Lindberg, *Beginnings*, 228–31; James A. Weisheipl ed., *Albertus Magnus and the Sciences: Commemorative Essays 1980* (Toronto: Pontifical Institute of Medieval Studies, 1980); Irvin M. Resnick, ed., *A Companion to Albertus the Great: Theology, Philosophy and the Sciences* (Leiden: Brill, 2013).

8. On the place of *De mineralibus* in Albertus's overall scheme of natural knowledge, see Albertus Magnus, *Book of Minerals*, trans. Dorothy Wyckoff (Oxford: Clarendon, 1967), xxvi–xxx.

9. On the authorship of these texts, see Lynn Thorndike, "Further Consideration of the *Experimenta, Speculum Astronomiae*, and *De Secretis Mulierum* Ascribed to Albertus Magnus," *Speculum* 30 (1955): 413–43; Paola Zambelli, *The* Speculum astronomiae *and Its Enigma: Astrology, Theology and Science in Albertus Magnus and His Contemporaries* (Dordrecht: Kluwer, 1992), 1–42; Weill-Parot, *Les "images astrologiques,"* 28–32; Scott E. Hendrix, *How Albert the Great's* Speculum astronomiae *Was Interpreted and Used by Four Centuries of Readers: A Study in Late Medieval Medicine, Astronomy and Astrology* (Lewiston: Edward Mellen, 2010), 10–35.

10. Albertus Magnus, *De Physica*, in *D. Alberti Magni Rartisbonensis Episcopi: Ordinis Praedicatorum, Opera omnia*, ed. Auguste Borgnet, vol. 3 (Paris: Vivès, 1890). See also Lindberg, *Beginnings*, 228–30; Mulchahey, *"First the Bow Is Bent,"* 257–59; French and Cunningham, *Before Science*, 180.

11. "Unde sciendum, quod Augustino in his quae sunt de fide et moribus plusquam Philosophis credendum est, si dissentiunt. Sed si de medicina loqueretur, plus ego crederem Galeno, vel Hipocrati: et si de naturis rerum loquatur, credo Aristoteli plus vel alii experto in rerum naturis." Albertus Magnus, *Commentarii in secundum librum sententiarum*, d.13. C. a.2, in *D. Alberti Magni Ratisbonensis Episcopi: Ordinis Praedicatorum, Opera omnia*, ed. Auguste Borgnet, vol. 27 (Paris: Vivès, 1894).

12. Lindberg, *Beginnings*, 24.

13. See the essays in Jeremiah Hackett, ed., *Roger Bacon and the Sciences* (New York: Brill, 1997), esp. Hackett, "Roger Bacon on *Scientia experimentalis*," 277–315. On Bacon and alchemy, see Dorothy Waley Singer, "Alchemical Writings Attributed to Roger Bacon," *Speculum* 7 (1932): 80–86; Edmund Brehm, "Francis Bacon's Place in the History of Alchemy," *Ambix* 23 (1976): 53–58; William R. Newman, "The Philosopher's Egg: Theory and Practice in the Alchemy of Roger Bacon," *Micrologus* 3 (1995): 75–101; William R. Newman, "An Overview of Roger Bacon's Alchemy," in Hackett, *Roger Bacon*, 317–36; Agostino Paravicini Bagliani, "Ruggero Bacone e l'alchimia della lunga vita," in *Alchimia e medicina nel Medioevo*, ed. Chiara Crisciani and Agostino Paravicini Bagliani (Florence: Simsel Edizioni del Galluzo, 2003), 33–54.

14. Albertus Magnus, *De mineralibus*, 2.1.1, in *D. Alberti Magni Rartisbonensis Episcopi: Ordinis Praedicatorum, Opera omnia*, ed. Auguste Borgnet, vol. 5 (Paris: Vivés, 1890); c.f. Augustine, *De civitate Dei*, 21.4.

15. Augustine, *De doctrina christiana*, 2.20, in *Sancti Aurelii Augustini, Hipponensis Episcopi, Opera omnia*, ed. J.-P. Migne, vol. 3.1 (Paris: Garnier Fratres, 1887); see also his remarks on Porphyry's magic, *De civitate Dei*, 10.11. On the canonists, see chap. 1 of this book, 43–48, 50–53.

16. "Antiquorum enim sapientium scripturam de sigillis lapidum pauci sciunt, nec sciri potest nisi simul et astronomia et magica et necromantiae scientiae sciantur." Albertus Magnus, *De mineralibus*, 2.3.1.

17. Albertus Magnus, *De mineralibus*, 2.3.1–2.

18. "Est autem principium in ipsa scientia omnia quaecumquae fiunt a natura vel arte, moveri a virtutibus coelestibus primo." Albertus Magnus, *De mineralibus*, 2.3.3.

19. "Est enim in homine duplex principium operum." Albertus Magnus, *De mineralibus*, 2.3.3.

20. "Incipit voluntas tunc ad motus siderum et figuras inclinari." Albertus Magnus, *De mineralibus*, 2.3.3.

21. "Et mira per tales imagines operabantur." Albertus, *De mineralibus*, 2.3.3.

22. On Albertus and alchemy, see J. R. Partington, "Albertus Magnus on Alchemy," *Ambix* 1 (1937): 3–20; Pearl Kibre, "Alchemical Writings Attributed to Albertus Magnus," *Speculum* 17 (1942): 499–518; Pearl Kibre, "Albertus Magnus on Alchemy," in Weisheipl, *Albertus Magnus and the Sciences*, 187–202.

23. Newman, "Technology and Alchemical Debate," 425.

24. Newman, "Technology and Alchemical Debate," 431–32.

25. Albertus Magnus, *Speculum astronomiae*, in Zambelli, *The* Speculum astronomiae, 257–67.

26. Thomas Aquinas, *Scriptum super sententiis*, 1.q.1, in *Thomae Aquinatis Scriptum super libros Sententiarum Magistri Petri Lombardi Episcopi Parisiensis*, vol. 1, ed. P. Mandonnet (Paris: P. Lethielleux, 1929).

27. Aquinas, *Scriptum super sententiis*, Proemium. q.1.

28. "Sed haec vera fides repudiat, per quam Angelos de caelo cecidisse, et Daemones esse credimus, et ex subtilitate suae naturae multa posse quae nos non possumus; et ideo illi qui eos ad talia facienda inducunt, malefici vocantur." Aquinas, *Scriptum super sententiis*, 4. d.34. q.1. a.3. Co.

29. Aquinas, *Scriptum super sententiis*, 2 d.7 q.3 a.1. See also Newman, "Technology and Alchemical Debate," 427–30; Principe, *Secrets of Alchemy*, 58–62. Aquinas later changed his opinion on the possibility of alchemy, however. See Tarrant, "Between Aquinas and Eymerich," 214–18.

30. Jan A. Aertsen, "Aquinas's Philosophy in Its Historical Setting," in *Cambridge Companion to Aquinas*, ed. Norman Kretzman and Eleonore Stump (Cambridge: Cambridge University Press, 1993), 13.

31. French and Cunningham, *Before Science*, 187.

32. Thomas Aquinas, *Summa contra gentiles*, 1.1–9, in *Sancti Thomae de Aquinatis Opera omnia iussu edita Leonis XIII P.M.*, vol. 13 (Rome: Typis Riccardi Garroni, 1918).

33. Aquinas, *Summa contra gentiles*, in *Opera omnia*, 13; on the creation, see bk. 2.

34. Aquinas, *Summa contra gentiles*, 2.32–38, 79–81, in *Opera omnia*, 13.

35. Thomas Aquinas, *Summa contra gentiles*, 3, esp. quotes 64–113, in *Sancti Tho-*

mae de Aquinatis Opera omnia iussu edita Leonis XIII P.M., vol. 14 (Rome: Typis Riccardi Garroni, 1926).

36. "Quod quidem facile perpendi potest si modus operationis earum attendatur." Aquinas, *Summa contra gentiles*, 3.105, in *Opera omnia*, 14.

37. Rutkin, *Sapientia Astrologica*, 1:304–12, quote 311. See also Weill-Parot, *Les "images astrologiques,"* 248.

38. Hinnebusch, *History of the Dominican Order*, 2:26–29.

39. For a discussion of the historiography of Averroism, see Craig Martin, "Rethinking Averroism," *Intellectual History Review* 17 (2007): 3–19. See also Fernand Van Steenberghen, *Aristotle in the West: The Origins of Latin Aristotelianism*, trans. Leonard Johnson (Louvain: E. Nauwerlaerts, 1955), 227–28; Fernand Van Steenberghen, *Thomas Aquinas and Radical Aristotelianism* (Washington, DC: Catholic University of America Press, 1980), 105–9; Richard C. Dales, "The Origin of the Doctrine of Double Truth," *Viator* 15 (1984): 169–79.

40. John F. Wippel, "The Condemnations of 1270 and 1277 at Paris," *Journal of Medieval and Renaissance Studies* 8 (1977): 180.

41. On the University of Paris's disciplinary procedures, see J. M. M. H. Thijssen, *Censure and Heresy at the University of Paris, 1200–1400* (Philadelphia: University of Pennsylvania Press, 1998), 1–39.

42. Wippel, "Condemnations of 1270 and 1277," 177.

43. Wippel, "Condemnations of 1270 and 1277," 173–85.

44. Lindberg, *Beginnings*, 234–35, Wippel, "Condemnations of 1270 and 1277," 174–77.

45. Thomas Aquinas, *Summa theologica*, 1. q.1 a.6. ad.2, 1 .q.1 a.2. co, in *Sancti Thomae de Aquinatis Opera omnia iussu edita Leonis XIII P.M.*, vol. 4 (Rome: Typographia poliglotta S. C. de Progranda fide, 1888). See also Beretta, "Orthodoxie philosophique," 68–71.

46. "Quia non eligit ea quae sunt vere a Christo tradita, sed ea quae sibi propria mens suggerit." Thomas Aquinas, *Summa theologica*, 2.2 q.11 a.1 co, in *Sancti Thomae de Aquinatis Opera omnia iussu edita Leonis XIII P.M.*, vol. 8 (Rome: Typographia poliglotta S. C. de Progranda fide, 1895).

47. "Et ideo haeresis est infidelitatis species pertinens ad eos qui fidem Christi profitentur, sed eius dogmata corrumpunt." Aquinas, *Summa theologica*, 2.2 q.11 a.1 co, in *Opera omnia*, 8.

48. Aquinas, *Summa theologica*, 2.2 q.11 a.3, in *Opera omnia*, 8.

49. Aquinas, *Summa theologica*, 2.2 q.11 a.2 co, in *Opera omnia*, 8.

50. Aquinas, *Summa theologica*, 2.2 q.11 a.3, in *Opera omnia*, 8.

51. "Sic igitur superstitio est vitium religioni oppositum secundum excessum, non quia plus exhibeat in cultum divinum quam vera religio: sed quia exhibet cultum divinum vel cui non debet, vel eo modo quo non debet." Thomas Aquinas, *Summa theologica*, 2.2 q.92 a.1 co, in *Sancti Thomae de Aquinatis Opera omnia iussu edita Leonis XIII P.M.*, vol. 9 (Rome: Typographia poliglotta S. C. de Progranda fide, 1897).

52. Aquinas, *Summa theologica*, 2.2 q. 94. a. 1 ad. 1, in *Opera omnia*, 9.

53. Aquinas, *Summa theologica*, 2.2 q. 90 a. 2, in *Opera omnia*, 9.

54. "Respondeo dicendum quod ars notoria et illicita est, et inefficax. Illicita quidem est, quia utitur quibusdam ad scientiam acquirendam quae non habent secundum se virtutem causandi scientiam: sicut inspectione quarundam figurarum, et

prolatione quorundam ignotorum verborum, et aliis huiusmodi." Aquinas, *Summa theologica*, 2.2 q.96 a.1 co., in *Opera omnia*, 9. On the *Ars notoria*, see chap. 1 of this book, 42–43.

55. Aquinas, *Summa theologica*, 2.2 q.96. a.2. in *Opera omnia*, 9.

56. "Res autem naturales habent quasdam virtutes occultas, quarum ratio ab homine assignari non potest: sicut quod adamas trahit ferrum, et multa alia quae Augustinus enumerat, XXI de Civ. Dei." Aquinas, *Summa theologica*, 2.2 q.96. a.2 arg.1, in *Opera omnia*, 9.

57. "Sed corpora naturalia sortiuntur quasdam virtutes occultas, speciem consequentes, ex impressione caelestium corporum. Ergo etiam corpora artificialia, puta imagines, sortiuntur aliquam virtutem occultam a corporibus caelestibus ad aliquos effectus causandos." Aquinas, *Summa theologica*, 2.2 q.96. a.2. arg.2, in *Opera omnia*, 9.

58. "Respondeo dicendum quod in his quae fiunt ad aliquos effectus corporales inducendos, considerandum est utrum naturaliter videantur posse tales effectus causare. Sic enim non erit illicitum: licet enim causas naturales adhibere ad proprios effectus." Aquinas, *Summa theologica*, 2.2. q.96 a.2 co, in *Opera omnia*, 9.

59. "Sic enim non erit illicitum: licet enim causas naturales adhibere ad proprios effectus. Si autem naturaliter non videantur posse tales effectus causare, consequens est quod non adhibeantur ad hos effectus causandos tanquam causae, sed solum quasi signa." Aquinas, *Summa theologica*, 2.2 q.96. a.2. co, in *Opera omnia*, 9.

60. "Si vero adiungantur vel characteres aliqui, vel aliqua nomina, vel aliae quaecumque variae observationes, quae manifestum est naturaliter efficaciam non habere, erit superstitiosum et illicitum." Aquinas, *Summa theologica*, 2.2 q. 96 a. 2 ad 1, in *Opera omnia*, 9.

61. Aquinas, *Summa theologica*, 2.2 q.96 a.2 ad.2, in *Opera omnia*, 9. Rutkin suggested that Aquinas may have taken this opportunity to "atone" for leaving open a loophole in the *Summa contra gentiles*. See *Sapientia Astrologica*, 1:311.

62. Aquinas, *Summa theologica*, 2.2 q.95 a.1–3, in *Opera omnia*, 9.

63. Aquinas, *Summa theologica*, 2.2 q.95 a.1, in *Opera omnia*, 9.

64. Aquinas, *Summa theologica*, 2.2 q.95 a.5, in *Opera omnia*, 9.

65. I follow here the reconstruction of events offered in Thijssen, *Censure and Heresy*, 40–56.

66. Wippel, "Condemnations of 1270 and 1277," 169–201; John F. Wippel, "Thomas Aquinas and the Condemnations of 1277," *Modern Schoolman* 72 (1995): 233–72; Edward Grant, "The Condemnation of 1277, God's Absolute Power and Physical Thought in the Late Middle Ages," *Viator* 10 (1979): 211–44; Thijssen, *Censure and Heresy*, 52–56.

67. "Condemnation of 219 Propositions," in *Medieval Political Philosophy*, ed. Ralph Lerner and Muhsin Mahdi, paperback ed. (Ithaca, NY: Cornell University Press, 1972), 335–56; Lindberg, *Beginnings*, 236–37.

68. "Condemnation of 219 Propositions," 338.

69. "*Summa de officio inquisitionis*," in *Quellen und Untersuchungen zur Geschichte de Hexenwahns und der Hexenverfolgung im Mittelalter*, ed. Joseph Hansen (Bonn: Carl Georgi, Universitäts-Buchdruckerei und Verlag, 1901), 43–44. Hansen gave a date of c. 1270 for this text; c.f. Boudet, who suggested c. 1260, *Entre science et* Nigromance, 557–58.

70. "*Summa de officio inquisitionis*," in Hansen, *Quellen*, 43–44.

Chapter Three

1. Traver, "Forging of an Intellectual Defense," 193.

2. Julien Théry-Astruc, "The Pioneer of Royal Theocracy: Guillaume de Nogaret and the Conflicts between Philip the Fair and the Papacy," in *The Capetian Century, 1214–1314*, ed. William Chester Jordan and Jenna Rebecca Phillips (Turnhout: Brepols, 2017), 223–26. On Boniface VIII, see E. Dupré Theseider, "Bonifacio VIII," in *Dizionario biografico degli italiani*, vol. 12 (Rome: Instituto della Enciclopedia Italiana, 1970), 146–70.

3. "*Super cathedram*," in Tanner, *Decrees*, 1:365–69. Traver, "Forging of an Intellectual Defense," 193. On the implications of *Super cathedram*, see Swanson, "Mendicant Problem," 217–20.

4. Alan Friedlander, *The Hammer of the Inquisitors: Brother Bernard Délicieux and the Struggle against the Inquisition in Fourteenth-Century France* (Leiden: Brill, 2000), 66–103.

5. Julien Théry, "A Heresy of State: Philip the Fair, the Trial of the 'Perfidious Templars' and the Pontificalization of the French Monarchy," *Journal of Medieval Religious Cultures* 39 (2013): 128–31; Théry-Astruc, "Pioneer of Royal Theocracy," 219–59. See also Malcolm Barber, *The Trial of the Templars* (Cambridge: Cambridge University Press, 1978).

6. P. N. R. Zutshi, "The Avignon Papacy," in *The New Cambridge Medieval History*, vol. 6, *c.1300–c.1415*, ed. Michael Jones (Cambridge: Cambridge University Press, 2000), 653–73; Joëlle Rollo-Koster, *Avignon and Its Papacy, 1309–1417: Popes, Institutions, and Society* (Lanham, MD: Rowman and Littlefield, 2016).

7. Barber, *Trial of the Templars*, 1–3; Théry, "Heresy of State," 128–32; Théry-Astruc, "Pioneer of Royal Theocracy," 247–49.

8. Council of Vienne, Decree 10, in Tanner, *Decrees*, 1:365–69.

9. Council of Vienne, Decree 26, in Tanner, *Decrees*, 1:380–82.

10. Council of Vienne, Canons 24, 28, and 29, in Tanner, *Decrees*, 1:379–80, 383–85. See Robin Vose's comments in *Dominicans, Muslims and Jews*, 134. His findings are supported by Maya Soifer Irish, who argued that anti-Jewish sentiment in Castile in the years leading up to the pogroms of 1391 was not systematically fomented by any institution within the Church. See her *Jews and Christians in Medieval Castile: Tradition, Coexistence and Change* (Washington, DC: Catholic University of America Press, 2016), 144–46; see also her account of the influence of Archdeacon Ferrán Martínez on the 1391 pogroms, "Towards 1391: The Anti-Judaic Preaching of Ferrán Martínez in Seville," in *The Medieval Roots of Antisemitism: Continuities and Discontinuities from the Middle Ages to the Present Day*, ed. Jonathan Adams and Cordelia Heß (New York: Routledge, 2018), 306–19.

11. "Sicut est circa aras idolorum nefarias preces emittere, sacrificia offerre, daemones consulere, eorum responsa suscipere. vel associant sibi propter sortes exercendas haereticos. vel faciant praedicta cum corpore vel sanguine Christi, vel in sortibus ut possint habere responsa, puerum rebaptizant. vel his similia." *Sexti decretalium*, 5.2.8.4, col. 622.

12. Kieckhefer, *European Witch Trials*, 10–15, 108–12. Kieckhefer dismissed the evidence supporting a supposed series of inquisitorial trials for sorcery in southern France; see 16–18.

13. Michael McVaugh, "Arnald of Villanova," in Gillespie, *Dictionary of Scientific Biography*, 1:289–91.

14. Arnald of Villanova, *De regimine sanitatis*, XI, in *Opera omnia* (Basel, 1585), col. 683–84.

15. Arnald of Villanova, *Antidotarium*, in *Opera omnia*, col. 391–92.

16. Klaassen, *Transformations of Magic*, 23.

17. On Pietro d'Abano's influence within the universities, see Nancy G. Siraisi, *Arts, Sciences at Padua: The Studium of Padua before 1350* (Toronto: Pontifical Institute of Medieval Studies, 1973), chap. 3; H. Darrel Rutkin, "Astrology," in *The Cambridge History of Science*, vol. 3, *Early Modern Science*, ed. Katherine Park and Lorraine Daston (Cambridge: Cambridge University Press 2006), 542–47; Hendrix, *Albert the Great's* Speculum astronomiae, 164–71; Jean-Patrice Boudet, Franck Collard, and Nicolas Weill-Parot, eds., *Médecine, astrologie et magie entre Moyen Âge et Renaissance: Autour de Pietro d'Abano* (Florence: Sismel Edizioni del Galluzzo, 2013).

18. Alain Boreau, *Satan the Heretic: The Birth of Demonology in the Medieval West*, trans. Teresa Lavender Fagan (Chicago: University of Chicago Press, 2006), 19–25; Rollo-Koster, *Avignon and Its Papacy*, 46–47; Rainer Decker, *Witchcraft and the Papacy: An Account Drawing on the Formerly Secret Records of the Roman Inquisition*, trans. H. C. Erik Midelfort (Charlottesville: University of Virginia Press, 2008), 24–29.

19. On Fournier, see Irene Bueno, *Defining Heresy: Inquisition, Theology and Papal Policy in the Time of Jacques Fournier*, trans. Isabella Bolognese, Tony Brophy, and Sarah Prodan (Leiden: Brill, 2015), esp. 289–95.

20. "Maleficos infectores gregis dominici effugare de medio domus dei." Cardinal Santa Sabina to Jean de Beaune and Bernard Gui, August 22, 1320, in Hansen, *Quellen*, 4.

21. On uses of the term *maleficus*, see Bailey, *Fearful Spirits*, 22.

22. Cardinal Santa Sabina to Jean de Beaune and Bernard Gui, August 22, 1320, in Hansen, *Quellen*, 4–5.

23. Kieckhefer, *European Witch Trials*, 10–15. See also Michael D. Bailey, "From Sorcery to Witchcraft: Clerical Conceptions of Magic in the Later Middle Ages," *Speculum* 76 (2001): 968.

24. Boreau, *Satan the Heretic*, 43–67; Bailey, *Fearful Spirits*, 79–80.

25. "Poenas omnes et singulas . . . quas de iure merentur heretici." *Super illius specula*, in Hansen, *Quellen*, 5.

26. Decker, *Witchcraft*, 29–31.

27. Bernard Gui, *Practica inquisitionis heretice pravitatis*, ed. C. Douais (Paris: Alphonse Picard, 1886). For further discussion of this manual, see Given, *Inquisition and Medieval Society*, 46; Sackville, "Inquisitor's Manual," 210; Derek Hill, *Inquisition in the Fourteenth Century: The Manuals of Bernard Gui and Nicholas Eymerich* (Woodbridge: York Medieval Press, 2019), 30–46.

28. Hayim Yerushalmi, "The Inquisition and the Jews of France in the Time of Bernard Gui," *Harvard Theological Review* 63 (1970): 317–76; Benjamin Z. Kedar, "Canon Law and the Burning of the Talmud," *Bulletin of Medieval Canon Law* (1979): 79–82; Cohen, *Friars and the Jews*, 51–99; Stow, "Church and the Jews," 204–19.

29. Mulchahey, *"First the Bow Is Bent,"* 378–84; on Aquinas's influence, see Hinnebusch, *History of the Dominican Order*, 1:149–90.

30. Gui, *Practica*, 5.6.2, 292–93.

31. Gui, *Practica*, 5.6.2, 292. For further discussion, see Bailey, "From Sorcery to Witchcraft," 969.

32. "Item, inquiratur maxime de hiis que sapiunt superstitionem quamcumque, aut irreverantiam, aut injuriam circa sacramenta Ecclesie et maxime circa sacramentum Corporis Christi, necnon circa cultum divinum et loca sacra." Gui, *Practica*, 5.6.2, 292–93.

33. Gui, *Practica*, 5.6.2, 293.

34. "Modus abjurandi pestem et errorem sortilegiorum aut divinationum et invocationum demonum, maxime ubi saperet heresim contra veritatem ac pietatem sacramenti Eucharistie vel baptismi seu aliorum sacramentorum, aut ubi in invocatione demonum exhiberetur aut fieret sacrificium vel immolatio demoni aut aliquid aliud errorem expressum continens contra fidem." Gui, *Practica*, 5.7.12, 301.

35. "Item, abjuro generaliter omnia sortilegia dampnata, maxime que ordinatur ad quoscumque effectus procurandos illicitos aut nocivos." Gui, *Practica*, 5.7.12, 301.

36. "Item, abjuro artem et modum faciendi ymagines de plumbo, seu de cera, seu de quacumque alia materia ad quoscumque effectus illicitos procurandos." Gui, *Practica*, 5.7.12, 301.

37. In addition to the two cases in Perugia and Modena discussed above, Kieckhefer identified a trial by a civil court in Salerno in 1371 and another heard by inquisitors in Como in 1360. He identified no other examples of the inquisition of magic. See *European Witch Trials*, 15–16. On the prosecution of magic in Tuscany, see Gene A. Brucker, "Sorcery in Early Renaissance Florence," *Studies in the Renaissance* 10 (1963): 8–9; Ugolino Nicolini, "La stregonaria a Perugia e in Umbria nel Medioevo: Con i testi di sette processi a Perugia e uno a Bologna," *Bolletino della Deputazione di Storia Patria per l'Umbria* 84 (1988): 30–38.

38. Nicolini, "La stregonaria a Perugia," 30–38.

39. "Richiestole come possa credere di far cosa meritoria invocando gli spiriti maligni, essendo cosa proibita dalla Chiesa e condannata come eresia." Grazia Biondi, *Benvenuta e l'Inquisitore: Un destino di donna nella Modena del '300* (Modena: Unione Donne Italiane—Centro documentazione donna, 1993); for the transcript of the trials translated into Italian, see 99–124, quote 105.

40. "Nonnulli etiam quandoque litterati, in hoc se opponunt, pretendentes id ad tuum non spectare officium secundum canonicas sanctiones." Gregory XI to Jacque de Morey (Jacobus de Morerio) OP, Inquisitor of France, August 14, 1374, in Hansen, *Quellen*, 15–16.

41. Baldini and Spruit, *Catholic Church and Modern Science*, 1:1983–84; Hill, *Inquisition in the Fourteenth Century*, 27–29; Karen Sullivan, *The Inner Lives of Medieval Inquisitors* (Chicago: University of Chicago Press, 2011), 170; Michael A. Ryan, *A Kingdom of Stargazers: Astrology and Authority in the Late Medieval Kingdom of Aragon* (Ithaca, NY: Cornell University Press, 2011), 126–27.

42. Hill, *Inquisition in the Fourteenth Century*, 193–99.

43. Nicholas Eymerich, *Directorium inquisitorum* (Rome, 1587), 2.42–43. For further discussion, see Hill, *Inquisition in the Fourteenth Century*, 47–57.

44. Eymerich, *Directorium*, 2.43, f.235.

45. Eymerich, *Directorium*, 2.43, ff.235–36.

46. Eymerich, *Directorium*, 2.43, ff. 236–39.

47. Eymerich, *Directorium*, 2.43, ff.236–40.

48. "Quidam sunt sortilegi et divinatores meri, sicut sunt qui agunt mere ex arte

chiromantiae, qui ex manus lineamentis divinant, et iudicant de effectibus naturalibus et conditionibus hominis." Eymerich, *Directorium*, 2.42, f.234.

49. "Quae quidem sapiunt haeresim manifeste." Eymerich, *Directorium*, 2.42, f.234.

50. "Admiscuerunt aliqua, quae sapiunt haeresim manifeste." Eymerich, *Directorium*, 2.42, f.234–35.

51. Eymerich, *Directorium*, 2.42, f.235.

52. See Tarrant, "Ambiguities of Censorship," 180–82.

53. For a further discussion of Aquinas's ideas, see chap. 2 of this book, 82–88, 91–100.

54. Eymerich did discuss astrology at greater length in a later text, *Contra astrologos imperitos*, completed in either 1395 or 1396. See Bailey, *Fearful Spirits*, 88–92.

55. "Nam quando non possunt pertingere ad finem intentum, daemonis auxilium quaerunt, invocant et imploránt et implorando obsecrant et sacrificant tacite, vel expresse." Eymerich, *Directorium*, 3, f.295.

56. Bailey, *Fearful Spirits*, 81–94; Tarrant, "Between Aquinas and Eymerich," 219–24.

57. Eymerich, *Directorium*, 2.23–29, ff.221–26. Baldini and Spruit, *Catholic Church and Modern Science*, 1:1983–84.

58. Kieckhefer, *European Witch Trials*, 18–23, 115–18; Duni, *Under the Devil's Spell*, 27–29.

59. Bruckner, "Sorcery in Early Renaissance Florence," 9–11, 21.

60. Bruckner, "Sorcery in Early Renaissance Florence," 13–20; for the original letter in Italian, see G. Sanesi, "Un episodio d'eresia nel 1383," *Bulletino Senese di Storia Patria* 3 (1896): 384–88.

61. Bruckner, "Sorcery in Early Renaissance Florence," 22–23.

62. "Sentenza del podestà di Milano contro Sibellia and Pierina," in Luisa Muraro, *La Signora del Gioco: Episodi della caccia alle streghe* (Milan: Feltrinelli, 1976), 240. For Kieckhefer's analysis of these trials, see "Mythologies of Witchcraft in the Fifteenth Century," *Magic, Ritual, and Witchcraft* 1 (2006): 96–98. On these trials and their possible roots in folk culture, see also Kieckhefer, *European Witch Trials*, 21–22; Gustav Henningsen, "'The Ladies from Outside': An Archaic Pattern of the Witches' Sabbath," in Ankarloo and Henningsen, *Early Modern European Witchcraft*, 203–4.

63. "Item dixisti quod illa domina docet vos de illa sotietate virtutes herbarum et etiam per signa que sibi hostenditis hostendit vobis omnia que sibi queritis de morbis et de furtis et de maleficiis." Muraro, *La Signora del Gioco*, 243.

64. "Quando ipsa voluit ire ad ipsum ludum vocavit spiritum Luciffellum qui semper venit ad eam et portavit eam ad ipsum ludum et cetera alias omni vice qua ipsum vocavit semper venit ad eam in forma hominis et secum locutus fuit et instruxit eam de omnibus que sibi petiit." Muraro, *La Signora del Gioco*, 244.

65. Kieckhefer, *European Witch Trials*, 22.

Chapter Four

1. Zutshi, "Avignon Papacy," 653–73; Rollo-Koster, *Avignon and Its Papacy*.

2. Howard Kaminsky, "The Great Schism," in Jones, *New Cambridge Medieval History*, 6:674–78; Oakley, *Conciliarist Tradition*, 32–37, 66–67.

3. Howard Kaminsky, *Simon de Cramaud and the Great Schism* (New Brunswick, NJ: Rutgers University Press, 1983), esp. 147–48; Kaminsky, "Great Schism," 686–

95; Anthony Black, "Popes and Councils," in *New Cambridge Medieval History*, vol. 7, *c.1415–c.1500*, ed. Christopher Allmand (Cambridge: Cambridge University Press, 1998), 65–66; Oakley, *Conciliarist Tradition*, 37, 64–65. For further discussion of the development Gerson's ideas, see John B. Morrall, *Gerson and the Great Schism* (Manchester: Manchester University Press, 1960); John J. Ryan, *The Apostolic Conciliarism of Jean Gerson* (Atlanta, GA: Scholars, 1998).

4. Kaminsky, "Great Schism," 695–96; A. Black, "Popes and Councils," 65–66; Oakley, *Conciliarist Tradition*, 37–38.

5. Council of Constance, Session 4, March 30, 1415, in Tanner, *Decrees*, 1:408–9.

6. Council of Constance, Session 39, July 28, 1417, in Tanner, *Decrees*, 1:438–43.

7. Council of Constance, Session 40, October 30, 1417, in Tanner, *Decrees*, 1:444–46; A. Black, "Popes and Councils," 66–67.

8. Council of Basel, Session 1, December 14, 1431, in Tanner, *Decrees*, 1:455–56; A. Black, "Popes and Councils," 69–70; Oakley, *Conciliarist Tradition*, 42–46.

9. A. Black, "Popes and Councils," 70; Oakley, *Conciliarist Tradition*, 46–47.

10. Council of Basel, Sessions 12–21, July 13, 1433–June 9, 1435, in Tanner, *Decrees*, 1:469–92; A. Black, "Popes and Councils," 70–71; Michael A. Mullett, *The Catholic Reformation* (London: Routledge, 1999), 3–4.

11. Council of Basel, Session 23, March 26, 1436, in Tanner, *Decrees*, 1:494–505; Oakley, *Conciliarist Tradition*, 49–50.

12. A. Black, "Popes and Councils," 71–72; Oakley, *Conciliarist Tradition*, 47–51.

13. A. Black, "Popes and Councils," 72–73; Oakley, *Conciliarist Tradition*, 47–51. On the Pragmatic Sanction and its implications, see Richard F. Costigan, "State Appointment of Bishops," *Journal of Church and State* 8 (1966): 87–88.

14. A. Black, "Popes and Councils," 70–73; Michael Mallet, "The Northern Italian States," in Allmand, *New Cambridge Medieval History*, 7:558–59; Bernhard Schimmelpfennig, *The Papacy*, trans. James Sievert (New York: Columbia University Press, 1992), 237–42; Mullet, *Catholic Reformation*, 11; David S. Peterson, "Religion and the Church," in *Italy in the Age of the Renaissance*, ed. John M. Najemy (Oxford: Oxford University Press, 2004), 71–76.

15. Thomas M. Izbicki, *Protector of the Faith: Cardinal Johannes de Turreremata and the Defence of the Institutional Church* (Washington, DC: Catholic University of America Press, 1981), 76–102.

16. Robson, *Franciscans*, 181–91; Andrews, *Other Friars*, 163–71.

17. Lino Leonardi, "The Bible in Italian," in *The New Cambridge History of the Bible*, vol. 2, *From 600–1450*, ed. Richard Marsden and E. Ann Matter (Cambridge: Cambridge University Press, 2012), 268–87; Peterson, "Religion and the Church," 59–81; Carol Everhart Quillen, "Humanism and the Lure of Antiquity," in Najemy, *Italy in the Age of the Renaissance*, 37–58.

18. Roberto Rusconi, "Manuali milanese di confessioni editi tra il 1474 e il 1523," *Archivium Franciscanum Historicum* 65 (1972): 110–11; Denys Hay, *The Church in Italy in the Fifteenth Century* (Cambridge: Cambridge University Press, 1977), 85; Robson, *Franciscans*, 181–91, 202–22; Andrews, *Other Friars*, 163–71; C. Robinson, "Cosimo de Medici and the Franciscan Observants at Bosci ai Frati," in *Cosimo il Vecchio*, ed. Ames Lewes (Oxford: Clarendon, 1992), 182–87.

19. Mullett, *Catholic Reformation*, 17–18.

20. Rusconi, *L'ordine dei peccati*, 246–60; Carolyn Muessig, "Bernardino da Siena and Preaching as a Vehicle for Religious Transformation," in Mixson and Roest,

Companion to Observant Reform, 185–203. See also Robert Black, "Education and the Emergence of a Literate Society," in Najemy, *Italy in the Age of the Renaissance*, 18–36.

21. For a general discussion of Observant reform and superstition, see Michael D. Bailey, "Reformers on Sorcery and Superstition," in Mixson and Roest, *Companion to Observant Reform*, 230–54. See also Paton, *Preaching Friars*, 264–306; Mormando, *Preacher's Demons*, 52–108, 164–218; Nirit Ben-Aryeh Debby, "Jews and Judaism in the Rhetoric of Popular Preachers: The Florentine Sermons of Giovanni Dominici (1345–1419) and Bernardino da Siena (1380–1444)," *Jewish History* 14 (2000): 175–200; Ben-Aryeh Debby, *Renaissance Florence*, 179–88. On the Monti di Pietà, see Vittorino Menenghin, *I Monti di Pietà in Italia dal 1462 al 1562* (Vicenza: L.I.E.F., 1986); Ariel Toaff, "Jews, Franciscans and the First 'Monti di Pietà' in Italy (1462–1500)," in McMichael and Myers, *Friars and Jews*, 239–54.

22. Kieckhefer, "Mythologies of Witchcraft," 88–89.

23. Domenico Mammoli, *The Record of the Trial and Condemnation of a Witch Matteuccia di Francesco, at Todi, 20 March 1428* (Rome: Cossidente, 1972), 13.

24. Kieckhefer, "Mythologies of Witchcraft," 88–89.

25. "Se sai d'alcuno indovino, o sognatore, o aguriatore, e dicono il vero, come dànno ad intendere di dire alcuna volta il vero, che il sanno per profezia ed eglino il sanno per incantazione diabolica, e se verrà vero quello che dicono, lo fanno per metterti in errore della fede cristiana e per farti adorare gl'idoli." Quoted in Nicolini, "La stregoneria a Perugia," 22.

26. "Peccant qui scribunt, qui portent, qui donant, qui vendunt, qui emunt, qui portanda docent, qui bona esse credunt." Quoted in Fabrizio Conti, "Preachers and Confessors against 'Superstitions': Bernardino Busti and Sermon 16 of His *Rosarium sermonum*," *Magic, Ritual and Witchcraft* 6 (2011): 78n69.

27. Kieckhefer, *European Witch Trials*, 118–21; Del Col, *L'Inquisizione*, 196.

28. Bernadette Paton, "To the Fire, to the Fire! Let Us Burn a Little Incense to God: Bernardino, Preaching Friars and Maleficio in Later Medieval Siena," in *No Gods Except Me: Orthodoxy and Religious Practice in Europe, 1200–1600*, ed. Charles Zika (Melbourne: History Department, University of Melbourne, 1991), 8–36.

29. Michael D. Bailey, *Battling Demons: Witchcraft, Heresy, and Reform in the Late Middle Ages* (University Park: Pennsylvania State University Press, 2003), 37.

30. "Imagines de cera aut alia re cum invocationibus etiam huiusmodi factis baptizant vel ordinant baptizari." Pope Eugenius IV to inquisitors, 1437, in Hansen, *Quellen*, 17–18.

31. "In veritate non incantaverit nec incantare sciret." Nicolini, "La stregoneria a Perugia," 42.

32. "Finsit et dolose demostravit scire incantare modo supradicto cum in veritate nichil sciret." Nicolini, "La stregoneria a Perugia," 43.

33. "Accessit ad quendam sacerdotum, cuius nomen pro meliori tacetur, et ab ipso suptili ingenio perquisivit habere et habuit quandam hostiam consecratam et in ipsa hostia ipsa Kacterina scribi fecit quasdam karacteras et certa nomina daemonum." Nicolini, "La stregoneria a Perugia," 45–50, quote 46.

34. "Multa maleficia fecerat cum corpore Christi a quodam presbitero suo amico consecrato." Nicolini, "La stregoneria a Perugia," 51.

35. Nicolini, "La stregoneria a Perugia," 53–54; Kieckhefer, "Mythologies of Witchcraft," 87, 96.

36. Bailey, *Battling Demons*, esp. 29–53, 91–138.

37. Johannes Nider, *Praeceptorium divinae legis, sive expositione decalogi* (Paris, 1474), 1.11. See also Bailey, *Battling Demons*, 40; Bailey, *Fearful Spirits*, 182–83.

38. On the revival of interest in the person of Aquinas and his thought during this period, see Paul Oskar Kristeller, "Thomism and the Italian Thought of the Renaissance," in *Medieval Aspects of Renaissance Learning: Three Essays*, ed. and trans. Edward P. Mahoney (Durham, NC: Duke University Press, 1974), 29–91; John W. O'Malley, "Some Renaissance Panegyrics of Aquinas," *Renaissance Quarterly* 27 (1974): 174–92; John W. O'Malley, *Praise and Blame in Renaissance Rome: Rhetoric, Doctrine and Reform in the Sacred Orators of the Papal Court, c. 1450–1521* (Durham, NC: Duke University Press, 1979); John W. O'Malley, "The Feast of Thomas Aquinas in Renaissance Rome: A Neglected Document and Its Import," *Rivista di storia della chiesa in Italia* 35 (1981): 1–27; E. P. Mahoney, "Saint Thomas Aquinas and the School of Padua at the End of the Fifteenth Century," *Proceedings of the American Catholic Philosophical Association*, vol. 48, *Thomas and Bonaventure, a Septicentenary Commemoration* (Washington, DC: Catholic University of America Press, 1974), 277–85.

39. "Quis aut occulta, aut futura investigare nititur per pacta aut occulta, aut manifesta inita cum demonibus." Nider, *Praeceptorium*, 1.11.A.

40. "Dicit tamen Thomas parte prima.quaestione.cxiiii.ar.iiii. quod omnes transmutations corporalium rerum, qui possunt fieri per aliquas virtutes, ad quas pertinent semina qui in elementis huius mundi inveniunt possunt fieri per operations daemonum." Nider, *Praeceptorium*, 1.11.7–8.

41. "Certitudinaliter possunt sciri per astronomos qui necessario eveniunt per cursus c[a]elorum." Nider, *Praeceptorium*, 1.11.38 (this edition of the text incorrectly marks this question as "Question 33").

42. "Non habet causam, et praecipue naturalem." Nider, *Praeceptorium*, 1.11.38.

43. Nider, *Praeceptorium*, 1.11.38. On Aquinas's idea, see chap. 2 of this book, 82–88, 91–100.

44. Kieckhefer, "Mythologies of Witchcraft," 82–87, 100–103, quote 86.

45. Kieckhefer, "Mythologies of Witchcraft," 98–103. For a useful summary of research into the Lausanne witch hunts, see Kathrin Utz Tremp, "Witches' Brooms and Magic Ointments: Twenty Years of Witchcraft Research at the University of Lausanne (1989–2009)," *Magic, Ritual, and Witchcraft* 5 (2010): 173–87.

46. Del Col, *L'Inquisizione*, 197–98; Tavuzzi, *Renaissance Inquisitors*, 252–55 (on the trial of Faelli, see 181); Stephen D. Bowd, "'Honeyed Flies' and 'Sugared Rats': Witchcraft, Heresy and Superstition in Bresciano, 1454–1535," in *The Religion of Fools? Superstition Past and Present*, ed. S. A. Smith and Alan Knight (Oxford: Oxford University Press, 2008), 142.

47. Del Col, *L'Inquisizione*, 197–98; Tavuzzi, *Renaissance Inquisitors*, 255; Bowd, "'Honeyed Flies' and 'Sugared Rats,'" 144–45.

48. Del Col, *L'Inquisizione*, 195–200; Duni, *Under the Devil's Spell*, 29–30; Tavuzzi, *Renaissance Inquisitors*, 155–74.

49. Paton, "To the Fire," 14–27.

50. "*Summis desiderantes*," in Christopher S. Mackay, *The Hammer of Witches: A Complete Translation of the* Malleus maleficarum (Cambridge: Cambridge University Press, 2009), 71–74.

51. On Sprenger's role in the authorship of the *Malleus*, see Mackay, *Hammer of Witches*, 5–6.

52. "Approbation," in Mackay, *Hammer of Witches*, 74–80, quote 75; for further

discussion of its production, see 9–10. On the composition of the *Malleus*, see Hans Peter Broedel, *The* Malleus maleficarum *and the Construction of Witchcraft: Theology and Popular Belief* (Manchester: Manchester University Press, 2003).

53. Mackay, *Hammer of Witches*, 75.

54. Mackay, *Hammer of Witches*, 75–76.

55. Michael Tavuzzi, "Lorenzo Soleri da Sant'Agata OP (ob. ca. 1510): The 'Inquisitor Cumanus' of the *Malleus Maleficarum*—a Biographical Note," in Hoyer, *Praedicatores, Inquisitores I*, 407–20.

56. For a succinct summary, see Mackay, *Hammer of Witches*, 10–11.

57. Mackay, *Hammer of Witches*, 1.1.co, 92–93. For a discussion of Aquinas's position, see chap. 2 of this book, 79–82.

58. Mackay, *Hammer of Witches*, 1.1.co, 93.

59. Mackay, *Hammer of Witches*, 1.1.co, 94–101. For a discussion of these canons, see chap. 1 of this book, 45–47.

60. Mackay, *Hammer of Witches*, 1.2.ag.8, 108.

61. Mackay, *Hammer of Witches*, 1.2.ra.8, 118. For a more detailed discussion of Aquinas's ideas on images, see chap. 2 of this book, 87–88, 95–98; for the marginalization of his thought on these matters in subsequent inquisitorial practice, see chap. 3 of this book.

62. Mackay, *Hammer of Witches*, 1.5.co, 142–43.

63. Mackay, *Hammer of Witches*, 1.5.co, 143–44. For further discussion, see Tarrant, "Ambiguities of Censorship," 177.

64. Mackay, *Hammer of Witches*, 1.5.co, 143–44; Neil Tarrant, "Reconstructing Thomist Astrology: Robert Bellarmine and the Papal Bull *Coeli et terrae*," *Annals of Science* 77 (2020): 4–6. Augustine, *De civitate Dei*, 5.5; see also *De civitate Dei*, 5.4, and *De doctrina christiana*, 2.21, in which Augustine also draws on different aspects of this story.

65. Anthony D. Wright characterized the period in Western Christian history between the mid-fifteenth and the mid-eighteenth centuries as the "Augustinian Moment" and placed the pursuit of superstition and witchcraft in this context; see *The Counter-Reformation: Catholic Europe and the Non-Christian World* (Frome: Butler and Tanner, 1982), 1–83.

66. On the development of Florentine Platonism and its influence on religious thought, see James Hankins, *Humanism and Platonism in the Italian Renaissance* (Rome: Edizione di Storia e Letteratura, 1994); Amos Edelheit, *Ficino, Pico, and Savonarola: The Evolution of Humanist Theology, 1461/2–1498* (Leiden: Brill, 2008).

67. Charles H. Lohr, "Metaphysics," in *The Cambridge History of Renaissance Philosophy*, ed. Charles Schmitt and Quentin Skinner (Cambridge: Cambridge University Press, 1988), 578–84; S. A. Farmer, *Syncretism in the West: Pico's 900 Theses (1486)—the Evolution of Traditional Religious and Philosophical Systems* (Temple, AZ: Medieval and Renaissance Texts and Studies, 1998), esp. 1–58.

68. Charles Boer, trans., *Marsilio Ficino's* Book of Life, 4th printing (Woodstock, NY: Spring Publications, 1994).

69. Boer, *Book of Life*, 144–50.

70. Boer, *Book of Life*, 150–51. On this passage of the *Summa contra gentiles*, see chap. 2 of this book, 87–88; on Ficino's use of Aquinas's arguments, see Brian Copenhaver, "Scholastic Philosophy and Renaissance Magic in the *De Vita* of Marsilio Ficino," *Renaissance Quarterly* 37 (1984): 532–34; Klaassen, *Transformations of Magic*, 30.

71. See chap. 3 of this book.

72. Stefano dall'Aglio, *Savonarola and Savonarolism*, trans. John Gagné (Toronto: Centre for Reformation and Renaissance Studies, 2010), 13–21; Donald Weinstein, *Savonarola: The Rise and Fall of a Renaissance Prophet* (New Haven, CT: Yale University Press, 2011), 27–41, 64–71.

73. Dall'Aglio, *Savonarola and Savonarolism*, 23–44; Weinstein, *Savonarola*, 75–114; Carol Bresnahan Menning, "The Monte's 'Monte': The Early Supporters of Florence's Monte di Pietà," *Sixteenth Century Journal* 23 (1992): 661–76.

74. "Li huomini che sono in alto stato constituiti essere involti in questo errore dalli quali imparano errare li inferiori." Girolamo Savonarola, *Tractato contra li astrologi* (Florence, 1495), f.5r. For a discussion of Savonarola's developing attitude to astrology and its relationship to Pico's work, see Remo Catani, "Girolamo Savonarola and Astrology," *The Italianist* 18 (1998): 71–90.

75. Giovanni Pico della Mirandola, *Disputationes adversus astrologiam divinatricem*, ed. and trans. Eugenio Garin, 2 vols. (Florence: Valecchi, 1946–52), 3.1–6, 10–12. For a discussion of Pico's work and its implications, see Steven Vanden Broecke, *The Limits of Influence: Pico, Louvain, and the Crisis of Renaissance Astrology* (Leiden: Brill, 2003).

76. "Nelli nostri tempi nelli quali quasi tutto el mondo e involto in questa pestifera fallacia." Savonarola, *Tractato*, f.5r.

77. Robert Westman, *The Copernican Question: Prognostication, Skepticism, and Celestial Order* (Berkeley: University of California Press, 2011), 25–28; Ottavia Niccoli, *Prophecy and People in Renaissance Italy*, trans. L. G. Cochrance (Princeton, NJ: Princeton University Press, 1990), 3–29.

78. Savonarola, *Tractato*, f.5v.

79. "Pero che non si chiamano divinatori quelli che prenuntiano le cose le quali ordinamente procedono dalle cause naturale o sempre o quasi sempre perche questo e concesso a l'huomo e è cosa humana: ma quelli che sanza speciale illuminatione divina presúmano di prenuntiare le cose future che non hanno determinata causa naturale cioe che possono essere o non essere indifferentemente e maxime quelle che appartenghono al libero arbitrio si chiamano propriamente divinatori nelle scripture: perche tentano di fare quello che appartiene a dio solo: e contra questi parla la scriptura in molti luoghi." Savonarola, *Tractato*, f.6v.

80. Savonarola, *Tractato*, f.8r–9r.

81. "Lunga cosa saria adducere le parole de doctori catholici cosi hebrei greci e latini come e d'altre regione antique e moderni: li quali sanza discrepantia in questo acordano tucti e saria dire quello medesimmo per altre parole." Savonarola, *Tractato*, f.9r. Savonarola was referring to *De doctrina christiana*, 2.21.

82. "Sancto Thomaso ancora nella secunda secunde q.lxxxxv.ar.v. e in molti altri luoghi dice che quelli che per la consideratione delle stelle cercano di cognoscere le cose future casuale e fortuite e maxime le operationi delli houmini future sono vani e superstitiosi." Savonarola, *Tractato*, f.8v.

83. On Aquinas and the doctrine of inclination, see chap. 2 of this book, 99–100.

84. "Et pero questa loro presumptione iustamente e anchora damnata dalli Canoni perche e molto nociua alla religione christiana attribuendo li mysterii della gratia alli cieli e faccendo li propheti astrologi: Et è tanta la temerita delli astrologi che etiam li miracoli e li martyrii de martyri che sono sopra ogni forza di natura gli vogliono attribuire al cielo." Savonarola, *Tractato*, f.10r.

85. "Certo contra di quelli che dicono simili cose non e da disputare altrimenti che

col fuoco." Savonarola, *Tractato*, f.10r. On Averroism in the university, see Grendler, *Universities of the Italian Renaissance*, 281–93.

86. Del Col, *L'Inquisizione*, 198; Tavuzzi, *Renaissance Inquisitors*, 255.

87. "Bernardina Stadera and the Demon in the Glass Spere," in Duni, *Under the Devil's Spell*, 80–84. On the "clerical underworld," see chap. 1 of this book, 42. For more detailed explorations of this phenomenon in late medieval and early modern Italy, see Duni, *Under the Devil's Spell*, 49–51; Tamar Herzig, "The Demons and the Friars: Illicit Magic and Mendicant Rivalry in Renaissance Bologna," *Renaissance Quarterly* 64 (2011): 1025–58.

88. For a discussion of the so-called doctrine of Averroism or double truth, see chap. 2 of this book, 88–89. On the discussion of the immortality of the soul in the later fifteenth century, see Paul F. Grendler, "Intellectual Freedom in Italian Universities: The Controversy over the Immortality of the Soul," in *Le côntrole des idées a la renaissance*, ed. Jesús Martínez de Bujanda (Geneva: Librarie Droz, 1996), 31–48; Grendler, *Universities of the Italian Renaissance*, 281–83.

89. Grendler, *Universities of the Italian Renaissance*, 284–88, 353–57; John Monfasani, "Aristotelians, Platonists, and the Missing Ockhamists: Philosophical Liberty in Pre-Reformation Italy," *Renaissance Quarterly* 46 (1993): 247–76.

Chapter Five

1. Mullet, *Catholic Reformation*, 3–4.

2. Hay, *Church in Italy*, 49–109; Peterson, "Religion and the Church," 72–76.

3. James McConica, *Erasmus* (Oxford: Oxford University Press, 1991), 45–62; A. G. Dickens and Whitney R. D. Jones, *Erasmus the Reformer* (London: Methuen, 1994), 3–111. On the circulation of Erasmus's works in Italy, see Silvana Seidel Menchi, *Erasmo in Italia (1520–1580)* (Turin: Bollati Boringhieri, 1987).

4. John W. O'Malley, "Introduction," in *Collected Works of Erasmus*, vol. 66, *Spiritualia: Enchidrion, De contemptu mundi, De vidua christiana*, ed. John W. O'Malley (Toronto: University of Toronto Press, 1998), ix–li.

5. Peterson, "Religion and the Church," 64. For a more detailed account of the spread of vernacular Bibles in Italy, see Fragnito, *La bibbia al rogo*, 23–74.

6. Steven D. Bowd, *Reform before the Reformation: Vicenzo Querini and the Religious Renaissance in Italy* (Leiden: Brill, 2002); on Christian humanism, see 45. On Contarini, see Gigliola Fragnito, "Contarini, Gasparo," in *Dizionario Biografico degli Italiani*, vol. 28 (1983), 172–92; Elisabeth G. Gleason, *Gasparo Contarini: Venice, Rome and Reform* (Berkeley: University of California Press, 1993); Constance M. Furey, *Erasmus, Contarini, and the Religious Republic of Letters* (New York: Cambridge University Press, 2006).

7. Kenneth J. Jorgensen, "The Theatines," in *Religious Orders of the Catholic Reformation: In Honor of John C. Olin on His Seventy-Fifth Birthday*, ed. Richard L. DeMolen (New York: Fordham University Press, 1994), 1–29.

8. Erika Rummel, *Jiménez de Cisneros: On the Threshold of Spain's Golden Age* (Tempe: Arizona Center for Medieval and Renaissance Studies, 1999), 18–28. On Lateran IV, see chap. 1 of this book, 56–57.

9. Rummel, *Jiménez de Cisneros*, 53–65.

10. Kieckhefer, *European Witch Trials*, 23.

11. For an overview, albeit one focused on the fourteenth century, see Robin Vose,

"Heresy Inquisitions in the Later Middle Ages," in Prudlo, *Companion to Heresy Inquisitions*, 141–71.

12. Henry Kamen, *The Spanish Inquisition: An Historical Revision* (London: Weidenfeld and Nicholson, 1997), 1–65; Helen Rawlings, *The Spanish Inquisition* (Malden, MA: Blackwell, 2006), 21–46, 47–71; Helen Rawlings, *Church, Religion and Society in Early Modern Spain* (Houndsmills: Palgrave Macmillan, 2002), 2–13. See also Gustav Henningsen, "The Archives and Historiography of the Spanish Inquisition," in Henningsen and Tedeschi, *Inquisition in Early Modern Europe*, 54–78.

13. Kamen, *Spanish Inquisition*, 103.

14. William Monter, *Frontiers of Heresy: The Spanish Inquisition from the Basque Lands to Sicily* (Cambridge: Cambridge University Press, 1990), 255–62; Rawlings, *Spanish Inquisition*, 125–30. On Eymerich, see chap. 3 of this book, 127–36.

15. C. Black, *Italian Inquisition*, 48–51, app., table D, 264.

16. Herzig, "Bridging North and South," 371–75.

17. Del Col, *L'Inquisizione*, 199; Tavuzzi, *Renaissance Inquisitors*, 255–56.

18. "Duabus mulieribus stregis maleficis ac diabolicis." Nicolini, "La stregioneria a Perugia," 66–73, quote 67.

19. Conti, "Preachers and Confessors," 62–91. For further discussion of the *Rosarium* and its context, see Fabrizio Conti, *Witchcraft, Superstition, and Observant Franciscan Preachers: Pastoral Approach and Intellectual Debate in Renaissance Milan* (Turnhout: Brepols, 2015), 53–88; Paton, "To the Fire," 24–25.

20. Conti, "Preachers and Confessors," 74–79.

21. Tarrant, "Ambiguities of Censorship," 184–85.

22. John W. O'Malley, *Giles of Viterbo on Church and Reform* (Leiden: Brill, 1968).

23. Nelson Minnich, "Concepts of Reform Proposed at the Fifth Lateran Council (with New Appendices)," *Archivium Historiae Pontificae* 7 (1969): 163–251, 252–53, reprinted in Nelson H. Minnich, *The Fifth Lateran Council (1512–17): Studies on Its Membership, Diplomacy and Proposals for Reform* (Aldershot: Ashgate Variorum, 1993), essay 4; Francis Oakley, "Conciliarism at the Fifth Lateran Council?," *Church History* 41 (1972): 452; Oakley, *Conciliarist Tradition*, 60–140. On conciliarism's survival and significance in the Iberian peninsula, see Xavier Tubau, ed., "Situating Conciliarism in Early Modern Spanish Thought," special issue, *Renaissance and Reformation* 43 (2019): 9–166.

24. Minnich, "Concepts of Reform," reprinted in Minnich, *Fifth Lateran Council*, essay 4.

25. Nelson H. Minnich, "The Participants at the Fifth Lateran Council," *Archivum historiae pontificiae* 12 (1974): 157–206, reprinted in Minnich, *Fifth Lateran Council*, essay 1.

26. Giles of Viterbo, "Egidio da Viterbo's Address to the Fifth Lateran Council," in *The Catholic Reformation: Savonarola to Ignatius Loyola*, ed. John C. Olin (New York: Harper and Row: 1969), 44. On Giles and conciliarism, see O'Malley, *Giles of Viterbo*, 120–21.

27. As Bowd has observed, the work defies easy categorization, for it contained ideas that could be associated with either the Italian Reformation or the Counter-Reformation; see *Reform before the Reformation*, 144–45. This was not the only proposal for reform addressed to Leo X; see Nelson H. Minnich, "Session IX: Origins of Decree *Supernae dispositionis arbitrio* (1514)—the Reform Proposals (1513) of Stefano Taleazzi for the Fifth Lateran Council (1512–17)," *Annuarium Historiae Concilio-*

rum 27/28 (1995/96), reprinted in Nelson H. Minnich, *The Decrees of the Fifth Lateran Council (1512–17): Their Legitimacy, Origins, Contents and Implementation* (Abingdon: Routledge, 2016), n. 5.

28. Stephen M. Beall and John J. Schmitt, eds. and trans., *Libellus Addressed to Leo X, Supreme Pontiff by Blessed Paolo Giustiniani and Pietro Querini Hermits of Camaldoli* (Milwaukee, WI: Marquette University Press, 2016), 34–111.

29. Beall and Schmitt, *Libellus*, 145–55.

30. Beall and Schmitt, *Libellus*, 155–71.

31. Beall and Schmitt, *Libellus*, 171–81.

32. Niccoli, *Prophecy and People*; Westman, *Copernican Question*, 62–75; Monica Azzolini, *The Duke and the Stars: Astrology and Politics in Renaissance Milan* (Cambridge, MA: Harvard University Press, 2013).

33. Beall and Schmitt, *Libellus*, 175.

34. Beall and Schmitt, *Libellus*, 179.

35. Beall and Schmitt, *Libellus*, 173.

36. Beall and Schmitt, *Libellus*, 181.

37. Beall and Schmitt, *Libellus*, 179.

38. On the education of the secular and regular clergy, see Hay, *Church in Italy*, 49–71. On the ambiguous boundaries between orthodoxy and superstition in religious practice in the later medieval and early modern periods, see, for example, David Gentilcore, *From Bishop to Witch: The System of the Sacred in Early Modern Terra d'Otranto* (Manchester: Manchester University Press, 1992); Ruggiero, *Binding Passions*; Jacobson Schutte, *Aspiring Saints*.

39. Beall and Schmitt, *Libellus*, 187–203.

40. Beall and Schmitt, *Libellus*, 203–11, quote 209. On their broader proposal for reform of the religious orders, see Nelson H. Minnich, "Egidio da Viterbo, the Reform of Religious Orders, and the Fifth Lateran Council (1512–1517)," in *Egidio da Viterbo Cardinale Agostiniano tra Roma e l'Europa del Rinascimento: Atti del Convegno, Viterbo, 22–23 settembre 2012—Roma, 26–28 settembre 2012*, ed. Myriam Chiabò, Rocco Ronzani, and Angelo Maria Vitale (Rome: Centro Culturale Agostiniano and Roma nel Rinascimento, 2014), 226–29.

41. Beall and Schmitt, *Libellus*, 211–13.

42. Nelson H. Minnich, "The Healing of the Pisan Schism (1511–13) (with New appendices)," *Archivium Historiae Conciliorum* 16 (1984): 59–192, 193–197, reprinted in Minnich, *Fifth Lateran Council*, essay 2. On the Pragmatic Sanction and its implications, see chap. 4 of this book, 148–49.

43. On the distinction between different forms of conciliarism, see Oakley "Conciliarism," 454; Minnich, "Egidio da Viterbo," 217–67.

44. Lateran V, Session 8, December 19, 1513, in Tanner, *Decrees*, 1:605. I follow here Eric Constant, "A Reinterpretation of the Fifth Lateran Council Decree *Apostolici regiminis* (1513)," *Sixteenth Century Journal* 33 (2002): 353–79. See also Grendler, *Universities of the Italian Renaissance*, 289–93. On the Augustinians' role in securing the decree, see O'Malley, *Giles of Viterbo*, 40–44.

45. Lateran V, Session 8, December 19, 1513, in Tanner, *Decrees*, 1:606.

46. O'Malley, *Giles of Viterbo*, 42–44.

47. Martin Pine, "Pomponazzi and the Problem of 'Double Truth,'" *Journal of the History of Ideas* 29 (1968): 163–76; Martin Pine, *Pietro Pomponazzi: Radical Philosopher of the Renaissance* (Padua: Antenore, 1986), 235–343.

48. O'Malley, *Giles of Viterbo*, 155–57; Minnich, "Egidio da Viterbo," 230–67; Minnich, "The Proposals for an Episcopal College at Lateran V," in *Ecclesia militans. Studien zur Konzilien und Reformationsgesticht. Remigius Bäumer zum 70. Geburtstag gewidmet*, ed. Walter Brandmüller, Herbert Immenkötter, and Erwin Iserloh (Paderborn: Ferdinand Schöningh, 1: 213–32, reprinted in Nelson H. Minnich, *The Fifth Lateran Council (1512–17): Studies on its Membership, Diplomacy and Proposals for Reform* (Aldershot and Brookfield VT: Ashgate Variorum, 1993), 221–22.

49. Lateran V, Session 9, May 5, 1514, in Tanner, *Decrees*, 1:614–21.

50. Lateran V, Session 9, May 5, 1514, in Tanner, *Decrees*, 1:621–24.

51. Lateran V, Session 9, May 5, 1514, in Tanner, *Decrees*, 1:625.

52. Lateran V, Session 9, May 5, 1514, in Tanner, *Decrees*, 1:625.

53. Nelson H. Minnich, "Session X: Decree *Inter multiplices* on Credit Organization (1515)—the Decree *Inter multiplices* of Lateran V on *Montes pietatis*," *Annuarium Historiae Conciliorum* 38 (2006), reprinted in Minnich, *Decrees of the Fifth Lateran Council*, essay 6.

54. Lateran V, Session 10, May 4, 1515, in Tanner, *Decrees*, 1:625–27.

55. Lateran V, Session 10, May 4, 1515, in Tanner, *Decrees*, 1:632–33; Gigliola Fragnito, "The Central and Peripheral Organization of Censorship," in *Church, Censorship and Culture in Early Modern Italy*, ed. Gigliola Fragnito, trans. Adrian Belton (Cambridge: Cambridge University Press, 2001), 15; Frajese, *Nascita dell'Indice*, 16–35; Nelson H. Minnich, "Session X: Decree *Inter sollicitudines* on Preventive Censorship (1515)—the Fifth Lateran Council and Preventive Censorship of Printed Books," *Annali dell Scuola Normale di Pisa, Classe di Lettere e Filosofia, serie 5* 2, (2010), reprinted in Minnich, *Decrees of the Fifth Lateran Council*, essay 7.

56. Lateran V, Session 11, December 19, 1516, in Tanner, *Decrees*, 1:634–38.

57. On *Omnis utriusque sexus*, see chap. 1 of this book, 56.

58. Lateran V, Session 11, December 19, 1516, in Tanner, *Decrees*, 1:645–49, quote 649.

59. Lateran V, Session 11, December 19, 1516, in Tanner, *Decrees*, 1:640–45; Oakley, *Conciliarist Tradition*, 56–57.

60. Del Col, *L'Inquisizione*, 199; C. Black, *Italian Inquisition*, 7.

61. Herzig, "Bridging North and South," 362, 375.

62. Duni, *Under the Devil's Spell*, 42–43.

63. "Anastasia 'la Frappona,' the Enchantress," in Duni, *Under the Devil's Spell*, 99–108. See also Duni, *Tra religione e magia: Storia del prete modenese Guglielmo Campana (1460?–1541)* (Florence: Olschki, 1999), 35–38.

64. "A Fearsome Wizard-Priest, Don Giglielmo Campana," in Duni, *Under the Devil's Spell*, 85–94, esp. 85–86.

65. Duni, *Under the Devil's Spell*, 86–89.

66. Duni, *Under the Devil's Spell*, 86–92.

Chapter Six

1. For the details of Luther's protest, see, for example, Andrew Johnson, *The Protestant Reformation in Europe* (Harlow: Longman, 1991), 15–33; Euan Cameron, *The European Reformation* (Oxford: Clarendon, 1991), 99–110; Scott Hendrix, "Martin Luther, Reformer," in Hsia, *Cambridge History of Christianity*, 6:3–19.

2. Maria J. Rodriguez-Salgado, "The Habsburg-Valois Wars," in *The New Cambridge Modern History*, vol. 2, *The Reformation, 1520–1559*, ed. Geoffrey R. Elton,

2nd ed (Cambridge: Cambridge University Press: 1990), 377–384; O'Malley, *Trent*, 56–57.

3. John Jefferies Martin, "Religion, Renewal and Reform in the Sixteenth Century," in Marino, *Early Modern Italy*, 32–41.

4. Dermot Fenlon, *Heresy and Obedience in Tridentine Italy: Cardinal Pole and the Counter Reformation* (Cambridge: Cambridge University Press, 1972), 10–11.

5. Thomas Braby, "Emergence and Consolidation of Protestantism in the Holy Roman Empire," in Hsia, *Cambridge History of Christianity*, 6:20–24.

6. Dickens and Jones, *Erasmus*, 225–33; Rawlings, *Spanish Inquisition*, 90–101; Rawlings, *Church, Religion and Society*, 30–32; Christopher F. Black, *Church, Religion and Society in Early Modern Italy* (Houndsmills: Palgrave, 2004), 3–9.

7. On Valdés, see Massimo Firpo, *Tra alumbrados e "spirituali": Studi su Juan de Valdés e il valdensianismo nella crisi religiosa nell'Europa del '500* (Turin: Einaudi, 1990); Massimo Firpo, "The Italian Reformation and Juan de Valdes," *Sixteenth Century Journal* 27 (1996): 353–64.

8. John W. O'Malley, *The First Jesuits* (Cambridge, MA: Harvard University Press, 1993).

9. Philip McNair, *Peter Martyr in Italy: An Anatomy of Apostasy* (Oxford: Oxford University Press, 1967), 139–79; Camilla Russell, *Giulia Gonzaga and the Religious Controversies of Sixteenth-Century Italy* (Turnhout: Brepols, 2006), 43–90.

10. Historians have debated the use of this term. I follow here Camilla Russell's position that "allowing for their immense complexity and diversity, the *spirituali* nevertheless were sufficiently distinct to warrant the term's use in historical analysis." Camilla Russell, "Religious Reforming Currents in Sixteenth-Century Italy: The *Spirituali* and the Tridentine Debates over Church Reform," *Journal of Religious History* 38 (2014): 460n12.

11. J. Martin, "Religion, Renewal and Reform," 35–36; Fenlon, *Heresy and Obedience*, 1–44.

12. Del Col, *L'Inquisizione*, 286–91; C. Black, *Italian Inquisition*, 11–16.

13. Del Col, *L'Inquisizione*, 286–91; C. Black, *Italian Inquisition*, 11–16.

14. Tavuzzi, *Renaissance Inquisitors*, chap. 5; see esp. his comments on p. 149.

15. Del Col, *L'Inquisizione*, 176–216; Duni, *Under the Devil's Spell*, 41–75; Tavuzzi, *Renaissance Inquisitors*, 193–95, 258. Tavuzzi also suggested that up to a thousand witches were burned each year during the 1520s, but this figure is derived from a sixteenth-century source and seems improbable (257).

16. See, for example, Herzig, "Bridging North and South," 361; C. Black, *Italian Inquisition*, 231; Midelfort, "Witch Craze," 14–16.

17. For further discussion, see the introduction to this book, 18–19.

18. Tavuzzi, *Renaissance Inquisitors*, 196–97.

19. Tavuzzi, *Renaissance Inquisitors*, 257–58.

20. "A Country Healer: Brighento Brighenti of Maranello," in Duni, *Under the Devil's Spell*, 113–17. C.f. Midelfort, "Witch Craze," 16, who distinguishes between prosecutions for diabolic witchcraft and those for sorcery, with the latter accusation being less likely to prompt a "panic." On the claim that denying the reality of witchcraft was a heresy, see chap. 4 of this book, 169.

21. "A Witch at the Sabbat: Orsolina 'la Rossa,'" in Duni, *Under the Devil's Spell*, 118–24.

22. Del Col, *L'Inquisizione*, 198; C. Black, *Italian Inquisition*, 6–7. Bowd's dis-

cussion of the tensions between secular leaders and inquisitors in the Valcamonica in the late fifteenth and early sixteenth centuries provides an excellent example of how these matters could be approached; see "'Honeyed Flies' and 'Sugared Rats,'" 144–51.

23. Tarrant, "Ambiguities of Censorship," 185–86.

24. Oakley, *Conciliarist Tradition*, 58–59; Xavier Tubau, "Hispanic Conciliarism and the Imperial Politics of Reform on the Eve of the Council of Trent," *Renaissance Quarterly* 70 (2017): 897–934. Tubau makes the important point that the idea of unequivocal Spanish support for the papacy and the Counter-Reformation is a misleading historiographical construction (897).

25. For a succinct summary of these issues, see J. Martin, "Religion, Renewal and Reform," 32–45.

26. O'Malley, *Trent*, 62. For the text of the *Consilium de emendanda ecclesia*, see Olin, *Catholic Reformation*, 182–97. On Lateran V, see chap. 5 of this book.

27. O' Malley, *Trent*, 59–68; Del Col, *L'Inquisizione*, 291; Peter Matheson, *Cardinal Contarini at Regensburg* (Oxford: Clarendon, 1972); Gleason, *Gasparo Contarini*, 186–256.

28. Fenlon, *Heresy and Obedience*, 46–51, and on the *Beneficio di Christo*, 69–88; Thomas F. Mayer, *Reginald Pole: Prince and Prophet* (Cambridge: Cambridge University Press, 2000), 103–33.

29. O'Malley, *Trent*, 70–71.

30. Silvana Seidel Menchi has argued that the extent of the diffusion of Lutheran ideas in Italy was limited and the commitment to them often shallow; see "Italy," 183–84.

31. Prosperi, *Tribunali della coscienza*, 35–56; Del Col, *L'Inquisizione*, 291–94.

32. Prosperi, *Tribunali della coscienza*, 38–39; C. Black, *Italian Inquisition*, 16, 19–20; Bonora, "Takeover of the Roman Inquisition," 250.

33. On the developing structures of censorship, see Del Col, *L'Inquisizione*, 315; Fragnito, "Central and Peripheral Organization," 13–49; Frajese, *Nascita dell'Indice*, 52–53.

34. Prosperi, *Tribunali della coscienza*, 57–116; Fragnito, "Central and Peripheral Organization," 13–49; C. Black, *Italian Inquisition*, 19–55, quote 42. See also the essays in Katherine Aron-Beller and Christopher Black, eds., *The Roman Inquisition: Centre versus Peripheries* (Leiden: Brill, 2018), which illustrate the persistence of tensions between center and periphery.

35. Prosperi, *Tribunali della coscienza*, 44–46; Del Col, *L'Inquisizione*, 287–93; C. Black, *Italian Inquisition*, 16–20.

36. Fenlon, *Heresy and Obedience*, 69–88, quote 88.

37. O'Malley, *Trent*, 72–89; R. Po Hsia, *The World of Catholic Renewal, 1540–1770*, 2nd ed. (Cambridge: Cambridge University Press, 2005), 10–13.

38. Council of Trent, Session 4, April 8, 1546, in Tanner, *Decrees*, 2:663–65. See also O'Malley, *Trent*, 89–102. The essential study of the circulation of scripture in the vernacular remains Fragnito, *La bibbia al rogo*.

39. Council of Trent, Session 5, June 17, 1546, Tanner, *Decrees*, 2:665–70. On the residence debates, see Hubert Jedin, *A History of the Council of Trent*, vol. 2, *The First Sessions at Trent, 1545–47*, trans. Dom Ernest Graf (London: Thomas Nelson, 1961), 317–27; O'Malley, *Trent*, 100–102. On the discussion of the mendicant orders at Lateran V, see chap. 5 of this book, 211–12.

40. O'Malley, *Trent*, 102–16; Mayer, *Reginald Pole*, 151–53.

41. O'Malley, *Trent*, 112–13.

42. Council of Trent, Session 6, January 13, 1547, in Tanner, *Decrees*, 2:671–83; O'Malley, *Trent*, 112–18. On the residence debate, see Jedin, *History of the Council of Trent*, 2:317–45. On Pole's withdrawal, see Fenlon, *Heresy and Obedience*, 133–36; Mayer, *Reginald Pole*, 153–54.

43. Council of Trent, Session 7, March 3, 1547, in Tanner, *Decrees*, 2:684–89; Jedin, *History of the Council of Trent*, 2:345–69; O'Malley, *Trent*, 116–21.

44. O'Malley, *Trent*, 121–38.

45. Russell, "Religious Reforming Currents," 465.

46. C. Black, *Italian Inquisition*, 20; Prosperi, *Tribunali della coscienza*, 117–34. On Paul III's broader efforts to restrain the Roman Inquisition, see M. Firpo, *La presa di potere*, 60–61. On the flight of Ochino and Vermigli, see McNair, *Peter Martyr in Italy*, 239–93.

47. R. Martin, *Witchcraft and the Inquisition*, app., table 1, "Witchcraft *processi* as compared to total *processi*, 1550–1650."

48. Duni, *Under the Devil's Spell*, 33.

Chapter Seven

1. Fenlon, *Heresy and Obedience*, 227–32; Frederic J. Baumgartner, "Henry II and the Papal Conclave of 1549," *Sixteenth Century Journal* 16 (1985): 301–14; O'Malley, *Trent*, 138–39; M. Firpo, *La presa di potere*, 3–51.

2. Council of Trent, Session 13, October 11, 1551, in Tanner, *Decrees*, 2:693–702; O'Malley, *Trent*, 146–48.

3. Council of Trent, Session 14, November 25, 1551, in Tanner, *Decrees*, 2:703–18; O'Malley, *Trent*, 145–54.

4. O'Malley, *Trent*, 154, 196. On this incident, see Xavier Tubau, "Between Ecclesiology and Diplomacy: Francisco de Vargas and the Council of Trent," *Renaissance and Reformation* 42 (2019): 117–21.

5. O'Malley, *Trent*, 154–60. On the *Varietas temporum*, see Hsia, *World*, 16–17; Mullett, *Catholic Reformation*, 53–54.

6. C. Black, *Italian Inquisition*, 22–23; M. Firpo, *La presa di potere*, 52–241; Del Col, *L'Inquisizione*, 305–15.

7. See Monter and Tedeschi, "Towards a Statistical Profile," 89–126; C. Black, *Italian Inquisition*, app., table A, 260–61.

8. Duni, *Under the Devil's Spell*, 33. For further discussion of this idea, see the introduction to this book, 14.

9. C. Black, *Italian Inquisition*, app., table A, 260–61, although c.f. the differing figures for Friuli supplied by Monter and Tedeschi, "Towards a Statistical Profile," app. 2, 145.

10. See Monter and Tedeschi, "Towards a Statistical Profile," app. 3, 107; C. Black, *Italian Inquisition*, app., table C, 263.

11. See C. Black, *Italian Inquisition*, app., tables A, C, and D, 260–61, 263–64. For Modena, see Duni, *Under the Devil's Spell*, 33.

12. Del Col's list of known witch trials overwhelmingly provides evidence from central and northern Italy. *L'Inquisizione*, 195–200. See also Monter and Tedeschi, "Towards a Statistical Profile," app. 3, 107.

13. For further discussion, see the introduction to this book, 12–13.

14. C. Black, *Italian Inquisition,* 160.

15. "Catalogue de 1544" and "Catalogue de 1551," in Jesús Martínez de Bujanda, ed., *Index des livres interdits,* vol. 1 (Sherbrooke: Éditions de l'Université de Sherbrooke, 1985), 466–72, 495–533; "Index de 1546," in de Bujanda, *Index des livres interdits,* 2:387–426. For the earlier Italian lists, see de Bujanda, *Index des livres interdits,* 3:378–93.

16. Paul F. Grendler, *The Roman Inquisition and the Venetian Press, 1540–1605* (Princeton, NJ: Princeton University Press, 1977), 94–95; Fausto Parente, "The Index, the Holy Office, the Condemnation of the Talmud and Publication of Clement VIII's Index," in Fragnito, *Church, Censorship and Culture,* 164.

17. "Index de Venise, 1554," in de Bujanda, *Index de livres interdits,* 3:394–423.

18. "Index de Venise, 1554," in de Bujanda, *Index des livres interdits,* 3:402.

19. de Bujanda, *Index des livres interdits,* 3:223, 350; Baldini and Spruit, *Catholic Church and Modern Science,* 1:1984–85. On Eymerich's list of banned works, see chap. 3 of this book, 136.

20. "Index de Venise, 1554," in de Bujanda, *Index des livres interdits,* 3:417.

21. See chap. 1 of this book, 53–57, and chap. 5, 214–15.

22. Kenneth R. Stow, "The Burning of the Talmud in 1553, in the Light of Sixteenth-Century Catholic Attitudes toward the Talmud," *Bibliothèque d'Humanisme et Renaissance* 34 (1972): 435–59; Fausto Parente, "La Chiesa e il 'Talmud': L'atteggiamento della Chiesa e del mondo cristiano nei confronti del Talmud e degli altri scritti rabbinici con particolare riguardo all'Italia tra XV e XVI secolo," in *Storia d'Italia,* vol. 11, *Gli ebrei in Italia, I: dall alto medio evo all'età di ghetti,* ed. C. Vivanti (Turin: Einaudi, 1996), esp. 580–98; Parente, "Index, Holy Office," 163–65. The Talmud was also burned in Venice. See Paul F. Grendler, "The Destruction of the Hebrew Books in Venice, 1568," *Proceedings of the American Academy for Jewish Research* 15 (1978): 103–7; Brian Pullan, *The Jews of Europe and the Inquisition of Venice, 1550–1670* (Oxford: Basil Blackwell, 1983), 82–84.

23. "Index de Venise, 1554," in de Bujanda, *Index des livres interdits,* 3:412. For further discussion of ritual magic, see chap. 1 of this book, 42–43; for Eymerich's condemnation of these arts, see chap. 3 of this book, 128–29.

24. "Index de Venise, 1554," in de Bujanda, *Index des livres interdits,* 3:394–423.

25. See chap. 3 of this book, 136.

26. Baldini and Spruit, *Catholic Church and Modern Science,* 1:136.

27. I have discussed the effects of this development in "Between Aquinas and Eymerich," 227–30, and "Ambiguities of Censorship," 186–95.

28. Mullett, *Catholic Reformation,* 53–54.

29. Mullett, *Catholic Reformation,* 54; H. Daniel-Rops, *The Catholic Reformation,* trans. John Warrington (London: J. M. Dent and Sons, 1962), 85–87.

30. M. Firpo, *Inquisizione Romana e Controriforma,* 261–358; C. Black, *Italian Inquisition,* 123–25; Del Col, *L'Inquisizione,* 293; Russell, *Giulia Gonzaga,* 151–72.

31. "Introduction historique," in de Bujanda, *Index des livres interdits,* 8:31–33; Frajese, *Nascita dell'Indice,* 70–78.

32. "Index de 1557," in de Bujanda, *Index des livres interdits,* 8:718–51. The practice of confining the Jews in ghettos originated in Venice in 1516, but Paul IV expanded the practice. For an overview, see Christopher F. Black, *Early Modern Italy: A Social History* (London: Routledge, 2001), 154–55.

33. "Magicae disciplinae, libri omnes & scripta." "Index de 1557," in de Bujanda, *Index des livres interdits,* 8:740.

34. "Libri omnes, et scripta Chyromantiae, Geomantiae, Hydromantiae, Physonomiae, Pyromantiae, vel Necromantiae, sive in quibus Sortilegia, veneficia, incantationes, Magicae Divinationes, vel Astrologica iudicia, circa Geneses, Nativitates, futuros eventus, sive particulares successus, status, vitae, vel mortis cuiusvis hominis describantur." "Index de 1557," in de Bujanda, *Index des livres interdits,* 8:737.

35. For a fuller discussion of Augustine and Aquinas's ideas of natural astrology, see chap. 2 of this book, 97–99, and on the fifteenth-century crisis of astrology, see chap. 4, 179–83.

36. Fragnito, "Central and Peripheral Organization," 16; Prosperi, *Tribunali della coscienza,* 228–43; C. Black, *Italian Inquisition,* 72–75.

37. "Index de 1559," in de Bujanda, *Index des livres interdits,* 8:754–801. Fragnito, "Central and Peripheral Organization," 15–16; Frajese, *Nascita dell'Indice,* 78–80.

38. "Introduction historique," in de Bujanda, *Index des livres interdits,* 8:38; Fragnito, *La bibbia al rogo,* 75–109, esp. 85.

39. "Index de 1559," in de Bujanda, *Index des livres interdits,* 8:758, 771.

40. "Libri omnes, et scripta Chiromantiae, Physionomiae, Aeromantiae, Geomantiae, Hydromantiae, Onomantiae, Pyromantiae, vel Necromantiae, sive in quibus Sortilegia, Veneficia, Auguria, Aruspicia, incantationes, Magicae artis, vel astrologiae iudiciariae Divinationes circa futuros contingentes eventus, aut eventuum successus sive fortuitos casus, iis tantum naturalibus observationibus exceptis, quae Navigationis, Agricolationis, sive Medicae artis iuvandae gratia conscripta sunt." "Index de 1559," in de Bujanda, *Index des livres interdits,* 8:775.

41. Ugo Baldini, "The Roman Inquisition's Condemnation of Astrology: Antecedents, Reasons and Consequences," in Fragnito, *Church, Censorship and Culture,* 90; Baldini and Spruit, *Catholic Church and Modern Science,* 1:452–53.

42. "Index de 1559," in de Bujanda, *Index des livres interdits,* 8:779 (on the identification of this work as the *Conciliator,* see 300).

43. "Anonymous, Censura of Antidotarium (Rome, post 1585, ante 1590)," in Baldini and Spruit, *Catholic Church and Modern Science,* 1:782–83.

44. Fragnito, *La bibbia al rogo,* 93; Fragnito, "Central and Peripheral Organization," 16–17; Frajese, *Nascita dell'Indice,* 80–87.

45. For further discussion of the *Instructio*'s comments on physiognomy, see Tarrant, "Ambiguities of Censorship," 189–90.

46. O'Malley, *Trent,* 162–63; Russell, *Giulia Gonzaga,* 174–82; C. Black, *Italian Inquisition,* 123–24. On the *sede vacante,* see Laurie Nussdorfer, "The Vacant See: Ritual and Protest in Early Modern Rome," *Sixteenth Century Journal* 18 (1987): 173–89 (on the disturbances in 1559, see 179–80); John M. Hunt, "Rome and the Vacant See," in *A Companion to Early Modern Rome, 1492–1692,* ed. Pamela H. Jones, Barbara Wisch, and Simon Ditchfield (Leiden: Brill, 2019), 99–114.

47. O'Malley, *Trent,* 164–67. See also Robert Knecht, *The French Civil Wars, 1562–98* (Harlow: Longman, 2000), 44–83; Mack P. Holt, *The French Wars of Religion, 1562–1629,* 2nd ed. (Cambridge: Cambridge University Press, 2005), 8–49.

48. O'Malley, *Trent,* 166–67.

49. O'Malley, *Trent,* 169–72.

50. O'Malley, *Trent,* 173–76.

51. O'Malley, *Trent*, 176–78; "Introduction historique," in de Bujanda, *Index des livres interdits*, 8:55–75.

52. O'Malley, *Trent*, 178–81; Hsia, *World*, 18–21. On the evangelical bishops' position during the third session of Trent, see Gigliola Fragnito, "La terza fase del concilio di Trento, Morone e gli 'Spirituali,'" in *Il Cardinal Giovanni Morone e l'ultima fase del Concilio di Trento*, ed. Massimo Firpo and Ottavia Niccoli (Bologna: Il Mulino, 2010), 53–78; Russell, "Religious Reforming Currents," 467–74.

53. O'Malley, *Trent*, 181–82.

54. Council of Trent, Session 21, July 16, 1562, and Session 22, September 17, 1562, in Tanner, *Decrees*, 2:726–32; O'Malley, *Trent*, 189–95.

55. On the preparation of the 1564 Index, see "Introduction historique," in de Bujanda, *Index des livres interdits*, 8:25–99 (for the composition of the commission, see 74; on the Rules, see 91–94). See also Fragnito, "Central and Peripheral Organization," 17; Fragnito, *La bibbia al rogo*, 98–109.

56. "Privo del officio e grado che tenea," "Per noi faria assai se si potesse trovare che il Direttorio fosse stato aprobato da qualche Papa, e se l'authore mori in riputazione di huomo da bene, pero V.S. vedi si potesse trovare qualche lume." "Dicono anche che il Lullo è tenuto in Spagno per Santo e che fa miracoli." "Camillo Campeggi to Girolamo Polizzi, Inquisitor of Cremona, in Rome (Trent, 19 October 1562)," in Baldini and Spruit, *Catholic Church and Modern Science*, 1:1991.

57. "Con le sue di V.S. delli 28. et 30. del passato ho ricevuto la condennatione del Talmud fatta da Innocenzo 4 et anche il testimonio di Lucemburgo contro il Lullo." "Camillo Campeggi to Girolamo Polizzi, Inquisitor of Cremona, in Rome (Trent, 16 November 1562)," in Baldini and Spruit, *Catholic Church and Modern Science*, 1:1992.

58. "Index de 1564," in de Bujanda, *Index des livres interdits*, 8:802–72. On the censorship of the Talmud, see Parente, "La Chiesa e il 'Talmud,'" 598–602; Parente, "Index, Holy Office," 168–69.

59. "Index de 1564," in de Bujanda, *Index des livres interdits*, 8:842, 852.

60. "Libri omnes, et scripta Geomantiae, Hydromantiae, Aeromantiae, Pyromantiae, Onomantiae, Chiromantiae, Necromantiae, sive in quibus continentur Sortilegia, Veneficia, Auguria, Auspicia, Incantationes artis Magicae, prorsus reiicuntur." "Index de 1564," in de Bujanda, *Index des livres interdits*, 8:818.

61. "Episcopi vero diligenter provideant, ne Astrologiae iudiciariae libri, tractatus, indices legantur, vel habeantur, qui de futuris contingentibus successibus, fortuitisue casibus, aut iis actionibus, quae ab humana voluntate pendent, certo aliquid eventurum affirmare audent. Permittuntur autem iudicia, et naturales observationes, quae navigationis, agriculturae, sive medicae artis iuvandae gratia conscripta sunt." "Index de 1564," in de Bujanda, *Index des livres interdits*, 8:818. The emphasis in the translation is mine.

62. For a discussion of this issue, see Baldini, "Roman Inquisition's Condemnation of Astrology," 90–91; Baldini and Spruit, *Catholic Church and Modern Science*, 1:455–56.

63. For further discussion, see Tarrant, "Reconstructing Thomist Astrology," 14–15.

64. O'Malley, *Trent*, 195–204, quote 197.

65. Council of Trent, Session 23, July 15, 1563, in Tanner, *Decrees*, 2:742–53; O'Malley, *Trent*, 205–23.

66. Council of Trent, Session 24, November 11, 1563, in Tanner, *Decrees*, 2:753–74; O'Malley, *Trent*, 223–42, quote 231.

67. Council of Trent, Session 25, December 3–4, 1563, in Tanner, *Decrees*, 2:774–99; O'Malley, *Trent*, 235–45.

68. Council of Trent, Session 25, December 3–4, 1563, in Tanner, *Decrees*, 2:774–99; Hsia, *World*, 22–23; O'Malley, *Trent*, 242–47.

69. O'Malley, *Trent*, 245–46.

Conclusion

1. Ugo Baldini and George V. Coyne, "The Louvain Lectures (Lectiones Lovanienses) of Bellarmine and the Autograph Copy of His 1616 Declaration to Galileo," *Vatican Observatory Publications, Special Series, Studi Galileani* 1 (1984); George V. Coyne and Ugo Baldini, "The Young Bellarmine's Thoughts on World Systems," in *The Galileo Affair: A Meeting of Faith and Science, Proceedings of the Cracow Conference, May 24–27, 1984*, ed. George V. Coyne (Vatican City: Specolo Vaticano, 1985), 103–9; Tarrant, "Reconstructing Thomist Astrology," 15–20.

2. "Autem magiam superstitiosam, et perniciosam docet." Robert Bellarmine, *Lectiones lovanienses*, Archivum Romanum Societatis Iesu, Opera Nostrorum 237, 393v.

3. William Eamon, "How to Read a Book of Secrets," in *Secrets and Knowledge in Medicine and Science, 1500–1800*, ed. Elaine Leong and Alisha Rankin (Farnham: Ashgate, 2011), 39.

4. William Eamon, *Science and the Secrets of Nature: Books of Secrets in Medieval and Early Modern Culture* (Princeton, NJ: Princeton University Press, 1994), esp. 194–233.

Bibliography

Reference Works

Dizionario biografico degli italiani. 92 vols. Rome: Instituto della Enciclopedia Italiana, 1960–.

Gillespie, Charles Coulston. *Dictionary of Scientific Biography.* 16 vols. New York: Charles Scribner's Sons, 1970–80.

Tedeschi, John. *The Italian Reformation of the Sixteenth Century and the Diffusion of Renaissance Culture: A Bibliography of the Secondary Literature (ca. 1750–1997).* In association with James M. Lattis. Modena: F.C. Panini, 2000.

Archival Sources

Bellarmine, Robert. *Lectiones lovanienses.* Archivium Romanum Societatis Iesu, Opera Nostrorum, 237.

Printed Primary Sources

Albertus Magnus. *Book of Minerals.* Translated by Dorothy Wyckoff. Oxford: Clarendon, 1967.

———. *D. Alberti Magni Rartisbonensis Episcopi: Ordinis Praedicatorum, Opera omnia.* Edited by Auguste Borgnet. 37 vols. Paris: Vivès, 1890–99.

Arnald of Villanova. *Opera omnia.* Basel, 1585.

Augustine. *Sancti Aurelii Augustini, Hipponensis Episcopi, Opera omnia.* 7 vols. Edited by J.-P. Migne. Paris: Garnier Fratres, 1841–87.

Baldini, Ugo, and Leen Spruit, eds. *Catholic Church and Modern Science: Documents from the Archives of the Holy Office and the Index.* Vol. 1, *Sixteenth-Century Documents.* Rome: Libreria Editrice Vaticana, 2009.

Barney, Steven A., W. J. Lewis, J. A. Beach, and Oliver Berghof, trans. *The Etymologies of Isidore of Seville.* Cambridge: Cambridge University Press, 2006.

Beall, Stephen M., and John J. Schmitt, eds. and trans. *Libellus Addressed to Leo X, Supreme Pontiff by Blessed Paolo Giustiniani and Pietro Querini Hermits of Camaldoli.* Milwaukee, WI: Marquette University Press, 2016.

Boer, Charles, trans. *Marsilio Ficino's* Book of Life. 4th printing. Woodstock, NY: Spring Publications, 1994.

Burchard of Worms. *Burchardi Vormatiensis Episcopi Opera Omnia.* Edited by J.-P. Migne. Paris: Garnier Fratres, 1880.

de Bujanda, Jesús Martínez, ed. *Index des livres interdits*. 10 vols. Sherbrook: Éditions de l'Université de Sherbrooke, 1980–1996.

Della Porta, Giambattista. *Magia naturalis*. Naples, 1558.

Eymerich, Nicholas. *Directorium inquisitorum*. Rome, 1587.

Farmer, S. A. *Syncretism in the West: Pico's 900 Theses (1486)—the Evolution of Traditional Religious and Philosophical Systems*. Temple, AZ: Medieval and Renaissance Texts and Studies, 1998.

Gratian. *Decretum gratiani*. Lyons, 1589.

Gui, Bernard. *Practica inquisitionis heretice pravitatis*. Edited by C. Douais. Paris: Alphonse Picard, 1886.

Hansen, Joseph, ed. *Quellen und Untersuchungen zur Geschichte de Hexenwahns und der Hexenverfolgung im Mittelalter*. Bonn: Carl Georgi, Universitäts-Buchdruckerei und Verlag, 1901.

Ivo of Chartres. *Sancti Ivonis Cartonensis episcopi opera omnia*. Edited by J.-P. Migne. Paris: J.-P. Migne, 1889.

Lerner, Ralph, and Muhsin Mahdi, eds. *Medieval Political Philosophy*. Paperback ed. Ithaca, NY: Cornell University Press, 1972.

Mackay, Christopher S. *The Hammer of Witches: A Complete Translation of the* Malleus maleficarum. Cambridge: Cambridge University Press, 2009.

Mammoli, Domenico. *The Record of the Trial and Condemnation of a Witch Matteuccia di Francesco, at Todi, 20 March 1428*. Rome: Cossidente, 1972.

Nicolini, Ugolino. "La stregonaria a Perugia e in Umbria nel Medioevo: Con i testi di sette processi a Perugia e uno a Bologna." *Bolletino della Deputazione di Storia Patria per l'Umbria* 84 (1988): 5–87.

Nider, Johannes. *Praeceptorium divinae legis, sive expositione decalogi*. Paris, 1474.

Olin, John C., ed. *The Catholic Reformation: Savonarola to Ignatius Loyola, Reform in the Church, 1495–1540*. New York: Harper and Row, 1969.

O'Malley, John W., ed. *Collected Works of Erasmus*. Vol. 66, *Spiritualia: Enchidrion, De contemptu mundi, De vidua christiana*. Toronto: University of Toronto Press, 1998.

Pico della Mirandola, Giovanni. *Disputationes adversus astrologiam divinatricem*. Edited and translated by Eugenio Garin. 2 vols. Florence: Valecchi, 1946–52.

Sanesi, G. "Un episodio d'eresia nel 1383." *Bulletino Senese di Storia Patria* 3 (1896): 384–88.

Savonarola, Girolamo. *Tractato contra li Astrologi*. Florence, 1495.

Sexti decretalium D. Bonifacii Papae VIII. Lyons, 1584.

Tanner, Norman, ed. *Decrees of the Ecumenical Councils*. 2 vols. London: Sheed and Ward, 1990.

Thomas Aquinas. *Sancti Thomae de Aquinatis Opera omnia iussu edita Leonis XIII P.M.* Vol. 4. Rome: Typographia poliglotta S. C. de Progranda fide, 1888.

———. *Sancti Thomae de Aquinatis Opera omnia iussu edita Leonis XIII P.M.* Vol. 8. Rome: Typographia poliglotta S. C. de Progranda fide, 1895.

———. *Sancti Thomae de Aquinatis Opera omnia iussu edita Leonis XIII P.M.* Vol. 9. Rome: Typographia poliglotta S. C. de Progranda fide, 1897.

———. *Sancti Thomae de Aquinatis Opera omnia iussu edita Leonis XIII P.M.* Vol. 13. Rome: Typis Riccardi Garroni, 1918.

———. *Sancti Thomae de Aquinatis Opera omnia iussu edita Leonis XIII P.M.* Vol. 14. Rome: Typis Riccardi Garroni, 1926.

———. *Thomae Aquinatis Scriptum super libros Sententiarum Magistri Petri Lombardi Episcopi Parisiensis*. Edited by P. Mandonnet. 4 vols. Paris: P. Lethielleux, 1929–47.

Valente, Michaela. "Della Porta e l'Inquisizione: Nuovi documenti dell'archivio del Sant'Uffizio." *Bruniana e campanelliana* 5 (1999): 415–34.

Wakefield, Walter L. *Heresy, Crusade and Inquisition in Southern France, 1100–1250*. London: George Allen and Unwin, 1974.

Wakefield, Walter L., and Austin P. Evans, eds. *Heresies of the High Middle Ages: Selected Sources Translated and Annotated*. New York: Columbia University Press, 1969.

Secondary Sources

Aertsen, Jan A. "Aquinas's Philosophy in Its Historical Setting." In *Cambridge Companion to Aquinas*, edited by Norman Kretzman and Eleonore Stump, 12–37. Cambridge: Cambridge University Press, 1993.

Andrews, Frances. *The Other Friars: The Carmelite, Augustinian and Pied Friars in the Middle Ages*. Woodbridge: Boydell, 2006.

Ansgar Kelly, Henry. "Canon Law and Chaucer on Licit and Illicit Magic." In *Law and the Illicit in Medieval Europe*, edited by Ruth Mazo Karras, Joel Kaye, and E. Ann Matter, 211–24. Philadelphia: University of Pennsylvania Press, 2008.

Arnold, John. *Inquisition and Power: Catharism and the Confessing Subject in Medieval Languedoc*. Philadelphia: University of Pennsylvania Press, 2001.

Aron-Beller, Katherine, and Christopher Black, eds. *The Roman Inquisition: Centre versus Peripheries*. Leiden: Brill, 2018.

Asztalos, Monika. "The Faculty of Theology." In de Ridder-Symoens, *History of the University*, 1:409–41.

Audisio, Gabriel. *The Waldensian Dissent: Persecution and Survival, c.1170–c.1570*. Cambridge: Cambridge University Press, 1999.

Azzara, Claudio. "The Papacy." In *Italy in the Early Middle Ages*, edited by Christina La Rocca, 109–15. Oxford: Oxford University Press, 2002.

Azzolini, Monica. *The Duke and the Stars: Astrology and Politics in Renaissance Milan*. Cambridge, MA: Harvard University Press, 2013.

Bagliani, Agostino Paravicini. "Ruggero Bacone e l'alchimia della lunga vita." In *Alchimia e medicina nel Medioevo*, edited by Chiara Crisciani and Agostino Paravicini Bagliani, 33–54. Florence: Simsel Edizioni del Galluzo, 2003.

Bailey, Michael D. *Battling Demons: Witchcraft, Heresy, and Reform in the Late Middle Ages*. University Park: Pennsylvania State University Press, 2003.

———. *Fearful Spirits, Reasoned Follies: The Boundaries of Superstition in Late Medieval Europe*. Ithaca, NY: Cornell University Press, 2013.

———. "From Sorcery to Witchcraft: Clerical Conceptions of Magic in the Later Middle Ages." *Speculum* 76 (2001): 960–90.

———. "Reformers on Sorcery and Superstition." In Mixson and Roest, *Companion to Observant Reform*, 230–54.

Baldini, Ugo. "The Roman Inquisition's Condemnation of Astrology: Antecedents, Reasons and Consequences." In Fragnito, *Church, Censorship and Culture*, 79–110.

Baldini, Ugo, and George V. Coyne. "The Louvain Lectures (Lectiones Lovanienses) of Bellarmine and the Autograph Copy of His 1616 Declaration to Galileo," *Vatican Observatory Publications, Special Series, Studi Galileani* 1 (1984).

Barber, Malcolm. *The Trial of the Templars*. Cambridge: Cambridge University Press, 1978.

Bartlett, Robert. *The Natural and the Supernatural in the Middle Ages*. Cambridge: Cambridge University Press, 2008.

Baumgartner, Frederic J. "Henry II and the Papal Conclave of 1549." *Sixteenth Century Journal* 16 (1985): 301–14.

Ben-Aryeh Debby, Nirit. "Jews and Judaism in the Rhetoric of Popular Preachers: The Florentine Sermons of Giovanni Dominici (1345–1419) and Bernardino da Siena (1380–1444)." *Jewish History* 14 (2000): 175–200.

———. *Renaissance Florence in the Rhetoric of Two Popular Preachers: Giovanni Dominici (1356–1419) and Bernardino da Siena (1380–1444)*. Turnhout: Brepols, 2001.

Beretta, Francesco. "Orthodoxie philosophique et Inquisition Romaine aux 16e–17e siècles: Un essai d'interpretation." *Historia philosophica* 3 (2005): 67–96.

Berman, Harold J. *Law and Revolution: The Formation of the Western Legal Tradition*. Cambridge, MA: Harvard University Press, 1983.

Biondi, Grazia. *Benvenuta e l'Inquisitore: Un destino di donna nella Modena del '300*. Modena: Unione Donne Italiane—Centro documentazione donna, 1993.

Black, Anthony. "Popes and Councils." In *The New Cambridge Medieval History*. Vol. 7, *c.1415–c.1500*, edited by Christopher Allmand, 65–86. Cambridge: Cambridge University Press, 1998.

Black, Christopher F. *Church, Religion and Society in Early Modern Italy*. Houndsmills: Palgrave, 2004.

———. *Early Modern Italy: A Social History*. London: Routledge, 2001.

———. *The Italian Inquisition*. New Haven, CT: Yale University Press, 2009.

Black, Robert. "Education and the Emergence of a Literate Society." In *Italy in the Age of the Renaissance*, edited by John M. Najemy, 18–36. Oxford: Oxford University Press, 2004.

Blumenthal, Uta-Renate. "The Papacy, 1024–1122." In *The New Cambridge Medieval History*. Vol. 4, pt. 2, *c.1024–c.1198*, edited by David Luscombe and Jonathan Riley-Smith, 8–37. Cambridge: Cambridge University Press, 2004.

Bolton, Brenda. "Papal Italy." In *Italy in the Central Middle Ages, 1000–1300*, edited by David Abulafia, 82–103. Oxford: Oxford University Press, 2004.

Bonora, Elena. "The Takeover of the Roman Inquisition." In *A Companion to Heresy Inquisitions*, edited by Donald S. Prudlo, 249–79. Leiden: Brill, 2019.

Boreau, Alain. *Satan the Heretic: The Birth of Demonology in the Medieval West*. Translated by Teresa Lavender Fagan. Chicago: University of Chicago Press, 2006.

Boudet, Jean-Patrice. *Entre science et* Nigromance: *Astrologie, divination et magie dans l'occident medieval (xiie–vxe siècle)*. Paris: Publications de la Sorbonne, 2006.

Boudet Jean-Patrice, Franck Collard, and Nicolas Weill-Parot, eds. *Médecine, astrologie et magie entre Moyen Âge et Renaissance: Autour de Pietro d'Abano*. Florence: Sismel Edizioni del Galluzzo, 2013.

Bowd, Stephen D. "'Honeyed Flies' and 'Sugared Rats': Witchcraft, Heresy and Superstition in Bresciano, 1454–1535." In *The Religion of Fools? Superstition Past and Present*, edited by S. A. Smith and Alan Knight, 134–56. Oxford: Oxford University Press, 2008.

———. *Reform before the Reformation: Vicenzo Querini and the Religious Renaissance in Italy*. Leiden: Brill, 2002.

Braby, Thomas. "Emergence and Consolidation of Protestantism in the Holy Roman

Empire." In *The Cambridge History of Christianity*. Vol. 6, *Reform and Expansion, 1500–1660*, edited by R. Po-chia Hsia, 20–36. Cambridge: Cambridge University Press, 2007.

Brambilla, Elena. *Alle origini del Sant' Uffizio: Penitenzia, confessione, e giustizia spirituale dal medioevo al XVI secolo*. Bologna: Il Mulino, 2000.

Brehm, Edmund. "Francis Bacon's Place in the History of Alchemy." *Ambix* 23 (1976): 53–58.

Bresnahan Menning, Carol. "The Monte's 'Monte': The Early Supporters of Florence's Monte di Pietà." *Sixteenth Century Journal* 23 (1992): 661–76.

Broedel, Hans Peter. *The* Malleus maleficarum *and the Construction of Witchcraft: Theology and Popular Belief*. Manchester: Manchester University Press, 2003.

Brooke, Rosalind B. *The Coming of the Friars*. London: George Allen and Unwin, 1975.

Bruckner, Gene A. "Sorcery in Early Renaissance Florence." *Studies in the Renaissance* 10 (1963): 7–24.

Bueno, Irene. *Defining Heresy: Inquisition, Theology and Papal Policy in the Time of Jacques Fournier*. Translated by Isabella Bolognese, Tony Brophy, and Sarah Prodan. Leiden: Brill, 2015.

Burnett, Charles. *Magic and Divination in the Middle Ages: Texts and Techniques in the Islamic and Christian Worlds*. Aldershot: Ashgate, 1996.

———. "Translation and Transmission of Greek and Arabic Science to Latin Christendom." In *The Cambridge History of Science*. Vol. 2, *Medieval Science*, edited by David C. Lindberg and Michael H. Shank, 341–64. Cambridge: Cambridge University Press, 2013.

———. "The Twelfth-Century Renaissance." In *The Cambridge History of Science*. Vol. 2, *Medieval Science*, edited by David C. Lindberg and Michael H. Shank, 365–84. Cambridge: Cambridge University Press, 2013.

Caldwell Ames, Christine. "Does Inquisition Belong to Religious History?" *American Historical Review* 110 (2005): 11–37.

———. *Righteous Persecution: Inquisition, Dominicans and Christianity in the Middle Ages*. Philadelphia: University of Pennsylvania Press, 2009.

Cameron, Euan. *The European Reformation*. Oxford: Clarendon, 1991.

———. *Waldenses: Rejections of Holy Church in Medieval Europe*. Oxford: Blackwell, 2000.

Catani, Remo. "Girolamo Savonarola and Astrology." *The Italianist* 18 (1998): 71–90.

Clark, Stuart. *Thinking with Demons: The Idea of Witchcraft in Early Modern Europe*. Oxford: Oxford University Press, 1997.

Clucas, Steven. "Exorcism, Conjuration and the Historiography of Early Modern Ritual Magic." In *Aisthetics of the Spirits: Spirits in Early Modern Science, Religion, Literature and Music*, edited by Steffen Schneider, 261–85. Göttingen: Vandenhoeck and Ruprecht, V&R Unipress, 2015.

Cohen, Jeremy. *The Friars and the Jews: The Evolution of Medieval Anti-Judaism*. Ithaca, NY: Cornell University Press, 1982.

Cohn, Norman. *Europe's Inner Demons: An Enquiry Inspired by the Great Witch-Hunt*. London: Heinemann for University of Sussex Press, 1975.

Colish, Marcia L. *Medieval Foundations of the Western Intellectual Tradition, 400–1400*. New Haven, CT: Yale University Press, 1997.

Collins, David J., ed. *The Cambridge History of Magic and Witchcraft in the West: From Antiquity to the Present*. Cambridge: Cambridge University Press, 2014.

———. "Learned Magic." In Collins, *Cambridge History of Magic*, 332–60.

———. "Scholasticism and High Medieval Opposition to Magic." In Page and Rider, *Routledge History of Medieval Magic*, 461–74.

Constant, Eric. "A Reinterpretation of the Fifth Lateran Council Decree *Apostolici regiminis* (1513)." *Sixteenth Century Journal* 33 (2002): 353–79.

Conti, Fabrizio. "Preachers and Confessors against 'Superstitions': Bernardino Busti and Sermon 16 of His *Rosarium sermonum*." *Magic, Ritual and Witchcraft* 6 (2011): 62–91.

———. *Witchcraft, Superstition, and Observant Franciscan Preachers: Pastoral Approach and Intellectual Debate in Renaissance Milan*. Turnhout: Brepols, 2015.

Copenhaver, Brian. "Scholastic Philosophy and Renaissance Magic in the *De Vita* of Marsilio Ficino." *Renaissance Quarterly* 37 (1984): 523–44.

Costigan, Richard F. "State Appointment of Bishops." *Journal of Church and State* 82 (1966): 82–96.

Cowdrey, H. E. J. "The Structure of the Church, 1024–1073." In *The New Cambridge Medieval History*. Vol. 4, pt. 1, *c.1024–c.1198*, edited by David Luscombe and Jonathan Riley-Smith, 229–67. Cambridge: Cambridge University Press, 2004.

Coyne, George V., and Ugo Baldini. "The Young Bellarmine's Thoughts on World Systems." In *The Galileo Affair: A Meeting of Faith and Science, Proceedings of the Cracow Conference, May 24–27, 1984*, edited by George V. Coyne, 103–9. Vatican City: Specolo Vaticano, 1985.

Crowley, Theodore. *Roger Bacon: The Problem of the Soul in His Philosophical Commentaries*. Louvain: Editions de l'Institut Supérior de Philosophie, 1950.

Cushing, Kathleen G. *Reform and the Papacy in the Eleventh Century: Spirituality and Social Change*. Manchester: Manchester University Press, 2005.

Dales, Richard C. "The Origin of the Doctrine of Double Truth." *Viator* 15 (1984): 169–79.

dall'Aglio, Stefano. *Savonarola and Savonarolism*. Translated by John Gagné. Toronto: Centre for Reformation and Renaissance Studies, 2010.

Daniel-Rops, H. *The Catholic Reformation*. Translated by John Warrington. London: J. M. Dent and Sons, 1962.

Decker, Rainer. *Witchcraft and the Papacy: An Account Drawing on the Formerly Secret Records of the Roman Inquisition*. Translated by H. C. Erik Midelfort. Charlottesville: University of Virginia Press, 2008.

Del Col, Andrea. *L'Inquisizione in Italia, dal XII al XXI secolo*. Milan: Mondadori, 2006.

de Ridder-Symoens, Hilde, ed. *A History of the University in Europe*. Vol. 1, *Universities in the Middle Ages*. Cambridge: Cambridge University Press, 1992.

Dickens, A. G., and Whitney R. D. Jones. *Erasmus the Reformer*. London: Methuen, 1994.

Ditchfield, Simon. "Of Dancing Cardinals and Mestizo Madonnas: Reconfiguring the History of Roman Catholicism in the Early Modern Period." *Journal of Early Modern History* 8 (2004): 386–408.

———. "Tridentine Catholicism." In *The Ashgate Companion to the Counter-Reformation*, edited by Alexandra Bamji, Geert H. Janseen, and Mary Laven, 15–31. Farnham: Ashgate, 2013.

Duni, Matteo. *Tra religione e magia: Storia del prete modenese Guglielmo Campana (1460?–1541)*. Florence: Olschki, 1999.

———. *Under the Devil's Spell: Witches, Sorcerers, and the Inquisition in Renaissance Italy*. Syracuse, NY: Syracuse University Press, 2007.

Eamon, William. "How to Read a Book of Secrets." In *Secrets and Knowledge in Medicine and Science, 1500–1800*, edited by Elaine Leong and Alisha Rankin, 23–46. Farnham: Ashgate, 2011.

———. *Science and the Secrets of Nature: Books of Secrets in Medieval and Early Modern Culture*. Princeton, NJ: Princeton University Press, 1994.

Edelheit, Amos. *Ficino, Pico, and Savonarola: The Evolution of Humanist Theology, 1461/2–1498*. Leiden: Brill, 2008.

Everhart Quillen, Carol. "Humanism and the Lure of Antiquity." In *Italy in the Age of the Renaissance*, edited by John M. Najemy, 37–58. Oxford: Oxford University Press, 2004.

Fanger, Claire, ed. *Conjuring Spirits: Texts and Traditions of Medieval Ritual Magic*. University Park: Pennsylvania State University Press, 1998.

———. "Introduction: Theurgy, Magic, and Mysticism." In *Invoking Angels: Theurgic Ideas and Practices, Thirteenth to Seventeenth Centuries*, edited by Claire Fanger, 1–33. University Park: Pennsylvania State University Press, 2012.

Fenlon, Dermot. *Heresy and Obedience in Tridentine Italy: Cardinal Pole and the Counter Reformation*. Cambridge: Cambridge University Press, 1972.

Findlen, Paula. "Science and Society." In *Early Modern Italy*, edited by John A. Marino, 166–87. Oxford: Oxford University Press, 2002.

Firpo, Luigi. "Filosofia italiana e controriforma." *Rivista di filosofia* 41 (1950): 150–73.

Firpo, Massimo. "Historiographical Introduction." In Tedeschi, *Italian Reformation*, xviii–xlix.

———. *Inquisizione Romana e Controriforma: Studi sul Cardinale Giovanni Morone (1509–1580) e il suo proceso di eresia*. 2nd edition. Brescia: La Morcelliana, 2005.

———. "The Italian Reformation and Juan de Valdes." *Sixteenth Century Journal* 27 (1996): 353–64.

———. *La presa di potere dell'Inquisizione Romana, 1550–1553*. Rome: Laterza, 2014.

———. *Tra alumbrados e "spirituali": Studi su Juan de Valdés e il valdensianismo nella crisi religiosa nell'Europa del '500*. Turin: Einaudi, 1990.

Flint, Valerie I. E. *The Rise of Magic in Medieval Europe*. Princeton, NJ: Princeton University Press, 1991.

Fragnito, Gigliola. "The Central and Peripheral Organization of Censorship." In Fragnito, *Church, Censorship and Culture*, 13–49.

———, ed. *Church, Censorship and Culture in Early Modern Italy*. Translated by Adrian Belton. Cambridge: Cambridge University Press, 2001.

———. *La bibbia al rogo: La censura ecclesiatica e i volgarizzamente della Scrittura (1471–1605)*. Bologna: Il Mulino, 1997.

Gigliola Fragnito. "La terza fase del concilio di Trento, Morone e gli 'Spirituali.'" In *Il Cardinal Giovanni Morone e l'ultima fase del Concilio di Trento*, edited by Massimo Firpo and Ottavia Niccoli, 53–78. Bologna: Il Mulino, 2010.

———. *Proibito capire: La chiesa e il volgare nella prima età moderna*. Bologna: Il Mulino, 2005.

———. *Rinascimento perduto: La letteratura italiana sotto gli occhi dei censori (secoli XV–XVII)*. Bologna: Il Mulino, 2019.

Frajese, Vittorio. *Nascita dell'Indice: La censura ecclesiatica dal Rinascimento alla Controriforma*. 2nd ed. Brescia: Morcelliana, 2008.

Frassetto, Michael. "Precursors to Religious Inquisitions: Anti-heretical Efforts to 1184." In *A Companion to Heresy Inquisitions*, edited by Donald S. Prudlo, 41–72. Leiden: Brill, 2019.

French, Roger, and Andrew Cunningham. *Before Science: The Invention of the Friars' Natural Philosophy*. Aldershot: Ashgate, 1996.

Friedlander, Alan. *The Hammer of the Inquisitors: Brother Bernard Délicieux and the Struggle against the Inquisition in Fourteenth-Century France*. Leiden: Brill, 2000.

Furey, Constance M. *Erasmus, Contarini, and the Religious Republic of Letters*. New York: Cambridge University Press, 2006.

Gentilcore, David. *From Bishop to Witch: The System of the Sacred in Early Modern Terra d'Otranto*. Manchester: Manchester University Press, 1992.

Ginzburg, Carlo. *The Night Battles: Witchcraft and Agrarian Cults in the Sixteenth and Seventeenth Centuries*. Translated by John Tedeschi and Anne C. Tedeschi. London: Routledge, 1983.

Given, James B. *Inquisition and Medieval Society: Power, Discipline, and Resistance in Languedoc*. Ithaca, NY: Cornell University Press, 1997.

Gleason, Elisabeth G. *Gasparo Contarini: Venice, Rome and Reform*. Berkeley: University of California Press, 1993.

Golinski, Jan. *Making Natural Knowledge: Constructivism and the History of Science*. Cambridge: Cambridge University Press, 1998.

Grant, Edward. "The Condemnation of 1277, God's Absolute Power and Physical Thought in the Late Middle Ages." *Viator* 10 (1979): 211–44.

———. *A History of Natural Philosophy: From the Ancient World to the Nineteenth Century*. Cambridge: Cambridge University Press, 2007.

Grendler, Paul F. "The Destruction of the Hebrew Books in Venice." *Proceedings of the American Academy for Jewish Research* 15 (1978): 103–30.

———. "Intellectual Freedom in Italian Universities: The Controversy over the Immortality of the Soul." In *Le côntrole des idées a la renaissance*, edited by Jesús Martínez de Bujanda, 31–48. Geneva: Librarie Droz, 1996.

———. *The Roman Inquisition and the Venetian Press, 1540–1605*. Princeton, NJ: Princeton University Press, 1977.

———. *The Universities of the Italian Renaissance*. Baltimore: Johns Hopkins University Press, 2002.

Hackett, Jeremiah, ed. *Roger Bacon and the Sciences*. New York: Brill, 1997.

———. "Roger Bacon on *Scientia experimentalis*." In Hackett, *Roger Bacon*, 277–315.

Hamilton, Bernard. "The Albigensian Crusade and Heresy." In *The New Cambridge Medieval History*. Vol. 5, *c.1198–c.1300*, edited by David Abulafia, 164–81. Cambridge: Cambridge University Press, 1999.

———. "Religion and the Laity." In *The New Medieval Cambridge History*. Vol. 4, pt. 1, *c.1024–c.1198*, edited by David Luscombe and Jonathan Riley-Smith, 499–533. Cambridge: Cambridge University Press, 2004.

Hankins, James. *Humanism and Platonism in the Italian Renaissance*. Rome: Edizione di Storia e Letteratura, 1994.

Haren, Michael. *The Western Intellectual Tradition from Antiquity to the Thirteenth Century*. Basingstoke: Macmillan, 1992.

Hay, Denys. *The Church in Italy in the Fifteenth Century*. Cambridge: Cambridge University Press, 1977.

Hegedus, Tim. *Early Christianity and Astrology*. New York: Peter Lang, 2007.

Hehl, Ernst-Dieter. "War, Peace and the Christian Order." In *The New Cambridge Medieval History*. Vol. 4, pt. 1, *c.1024–c.1198*, edited by David Luscombe and Jonathan Riley-Smith, 185–228. Cambridge: Cambridge University Press, 2004.

Hen, Yitzhak. "The Early Medieval West." In Collins, *Cambridge History of Magic*, 183–206.

Hendrix, Scott. "Martin Luther, Reformer." In *The Cambridge History of Christianity*. Vol. 6, *Reform and Expansion, 1500–1660*, edited by R. Po-chia Hsia, 3–19. Cambridge: Cambridge University Press, 2007.

Hendrix, Scott E. *How Albert the Great's* Speculum astronomiae *Was Interpreted and Used by Four Centuries of Readers: A Study in Late Medieval Medicine, Astronomy and Astrology*. Lewiston: Edward Mellen, 2010.

Henningsen, Gustav. "The Archives and Historiography of the Spanish Inquisition." In *The Inquisition in Early Modern Europe: Studies in Sources and Methods*, edited by Gustav Henningsen and John Tedeschi, 54–78. DeKalb: Northern Illinois University Press, 1986.

———. "'The Ladies from Outside': An Archaic Pattern of the Witches' Sabbath." In *Early Modern European Witchcraft: Centres and Peripheries*, edited by Bengt Ankaloo and Gustav Henningsen, 191–215. Oxford: Oxford University Press, 1990.

Henry, John. "The Fragmentation of Renaissance Occultism and the Decline of Magic." *History of Science* 46 (2008): 1–48.

Herzig, Tamar. "Bridging North and South: Inquisitorial Networks and Witchcraft Theory on the Eve of the Reformation." *Journal of Early Modern History* 12 (2008): 361–82.

———. "The Demons and the Friars: Illicit Magic and Mendicant Rivalry in Renaissance Bologna." *Renaissance Quarterly* 64 (2011): 1025–58.

Hill, Derek. *Inquisition in the Fourteenth Century: The Manuals of Bernard Gui and Nicholas Eymerich*. Woodbridge: York Medieval Press, 2019.

Hinnebusch, William A. *The History of the Dominican Order*. 2 vols. New York: Alba House, 1965–1973.

Holt, Mack P. *The French Wars of Religion, 1562–1629*. 2nd ed. Cambridge: Cambridge University Press, 2005.

Howe, John. *Before the Gregorian Reform: The Latin Church at the Turn of the First Millennium*. Ithaca, NY: Cornell University Press, 2016.

———. "Re-forging the 'Age of Iron,' Part 1: The Tenth Century as the End of the Ancient World?" *History Compass* 8 (2010): 866–87.

———. "Re-forging the Age of Iron, Part 2: The Tenth Century in a New Age?" *History Compass* 8 (2010): 1000–1022.

Hoyer, Wolfram, ed. *Praedicatores, Inquisitores I: The Dominicans and the Medieval Inquisition—Acts of the First International Seminar on the Dominicans and the Inquisition, Rome 23–25 February 2002*. Rome: Instituto Storico Domenico, 2004.

Hsia, R. Po. *The World of Catholic Renewal, 1540–1770*. 2nd ed. Cambridge: Cambridge University Press, 2005.

Hunt, John M. "Rome and the Vacant See." In *A Companion to Early Modern Rome, 1492–1692*, edited by Pamela H. Jones, Barbara Wisch, and Simon Ditchfield, 99–114. Leiden: Brill, 2019.

Irish, Maya Soifer. *Jews and Christians in Medieval Castile: Tradition, Coexistence and Change*. Washington, DC: Catholic University of America Press, 2016.

———. "Towards 1391: The Anti-Judaic Preaching of Ferrán Martínez in Seville."

In *The Medieval Roots of Antisemitism: Continuities and Discontinuities from the Middle Ages to the Present Day*, ed. Jonathan Adams and Cordelia Heß, 306–19. New York: Routledge, 2018.

Izbicki, Thomas M. *Protector of the Faith: Cardinal Johannes de Turrecremata and the Defence of the Institutional Church*. Washington, DC: Catholic University of America Press, 1981.

Jacobson Schutte, Anne. *Aspiring Saints: Pretense of Holiness, Inquisition, and Gender in the Republic of Venice, 1618–1750*. Baltimore: Johns Hopkins University Press, 2001.

———. "Religion, Spirituality, and the Post-Tridentine Church." In *Early Modern Italy*, edited by John A. Marino, 125–42. Oxford: Oxford University Press, 2002.

Jansen, Katherine. "The Word and Its Diffusion." In *The Cambridge History of Christianity*, vol. 4, *Christianity in Western Europe, c.1100–c.1500*, edited by Miri Rubin and Walter Simons, 114–32. Cambridge: Cambridge University Press, 2009.

Jedin, Hubert. *A History of the Council of Trent*. Vol. 2, *The First Sessions of Trent, 1545–47*. Translated by Dom Ernest Graf. London: Thomas Nelson, 1961.

Johnson, Andrew. *The Protestant Reformation in Europe*. Harlow: Longman, 1991.

Jorgensen, Kenneth J. "The Theatines." In *Religious Orders of the Catholic Reformation: In Honor of John C. Olin on His Seventy-Fifth Birthday*, edited by Richard L. DeMolen, 1–29. New York: Fordham University Press, 1994.

Kahlos, Maijastina. "The Early Church." In Collins, *Cambridge History of Magic*, 148–82.

Kamen, Henry. *The Spanish Inquisition: An Historical Revision*. London: Weidenfeld and Nicholson, 1997.

Kamerick, Kathleen. "Pastoral Literature and Preaching." In Page and Rider, *Routledge History of Medieval Magic*, 475–86.

Kaminsky, Howard. "The Great Schism." In *The New Cambridge Medieval History*. Vol. 6, *c.1300–c.1415*, edited by Michael Jones, 674–96. Cambridge: Cambridge University Press, 2000.

———. *Simon de Cramaud and the Great Schism*. New Brunswick, NJ: Rutgers University Press, 1983.

Kedar, Benjamin Z. "Canon Law and the Burning of the Talmud." *Bulletin of Medieval Canon Law* (1979): 79–82.

Kibre, Pearl. "Albertus Magnus on Alchemy." In Weisheipl, *Albertus Magnus and the Sciences*, 187–202.

———. "Alchemical Writings Attributed to Albertus Magnus." *Speculum* 17 (1942): 499–518.

Kieckhefer, Richard. *The European Witch Trials: Their Foundations in Popular and Learned Culture, 1300–1500*. London: Routledge, 1976.

———. *Magic in the Middle Ages*. 2nd ed. Cambridge: Cambridge University Press, 2014.

———. "Mythologies of Witchcraft in the Fifteenth Century." *Magic, Ritual, and Witchcraft* 1 (2006): 79–108.

———. "The Office of the Inquisition and Medieval Heresy: The Transition from Personal to Institutional Jurisdiction." *Journal of Ecclesiastical History* 46 (1995): 36–61.

Klaassen, Frank. "Necromancy." In Page and Rider, *Routledge History of Medieval Magic*, 201–211.

———. *The Transformations of Magic: Illicit Learned Magic in the Later Middle Ages and Renaissance.* University Park: Pennsylvania State University Press, 2013.

Knecht, Robert. *The French Civil Wars, 1562–98.* Harlow: Longman, 2000.

Kristeller, Paul Oskar. "Thomism and the Italian Thought of the Renaissance." In *Medieval Aspects of Renaissance Learning: Three Essays,* edited and translated by Edward P. Mahoney, 29–91. Durham, NC: Duke University Press, 1974.

Lambert, Malcolm. *Medieval Heresy: Popular Movements from the Gregorian Reform to the Reformation.* 3rd ed. Oxford: Blackwell, 2002.

Langmuir, Gavin. *Towards a Definition of Antisemitism.* Berkeley: University of California Press, 1990.

Laursen, John Christian, and Cary J. Nederman, eds. *Beyond the Persecuting Society: Religious Toleration before the Enlightenment.* Philadelphia: University of Pennsylvania Press, 1998.

Lawrence, C. H. *The Friars: The Impact of the Early Mendicant Movement on Western Society.* London: Longman, 1994.

Leonardi, Lino. "The Bible in Italian." In *The New Cambridge History of the Bible.* Vol. 2, *From 600–1450,* edited by Richard Marsden and E. Ann Matter, 268–87. Cambridge: Cambridge University Press, 2012.

Lindberg, David. *The Beginnings of Western Science: The European Scientific Tradition in Philosophical, Religious and Institutional Context, 600 B.C. to A.D. 1450.* Chicago: University of Chicago Press, 1992.

Lohr, Charles H. "The Medieval Interpretation of Aristotle." In *The Cambridge History of Later Medieval Philosophy: From the Rediscovery of Aristotle to the Disintegration of Scholasticism, 1100–1600,* edited by Norman Kretzman, Anthony Kenny, and Jan Pinborg, 80–98. Cambridge: Cambridge University Press, 1982.

———. "Metaphysics." In *The Cambridge History of Renaissance Philosophy,* edited by Charles Schmitt and Quentin Skinner, 537–638. Cambridge: Cambridge University Press, 1988.

Lotz-Heumann, Ute. "The Concept of 'Confessionalisation': A Historiographical Paradigm in Dispute." *Memoria y Civilización* 4 (2001): 93–114.

———. "Imposing Church and Social Discipline." In *The Cambridge History of Christianity.* Vol. 6, *Reform and Expansion, 1500–1600,* edited by R. Po-chia Hsia, 244–60. Cambridge: Cambridge University Press, 2007.

Mahoney, E. P. "Saint Thomas Aquinas and the School of Padua at the End of the Fifteenth Century." In *Proceedings of the American Catholic Philosophical Association.* Vol. 48, *Thomas and Bonaventure, a Septicentenary Commemoration,* 277–85. Washington, DC: Catholic University of America Press, 1974.

Mallet, Michael. "The Northern Italian States." In *The New Cambridge Medieval History.* Vol. 7, *c.1415–c.1500,* edited by Christopher Allmand, 547–70. Cambridge: Cambridge University Press, 1998.

Marrone, Steven P. "Magic and Natural Philosophy." In Page and Rider, *Routledge History of Medieval Magic,* 287–98.

Martin, Craig. "Rethinking Averroism." *Intellectual History Review* 17 (2007): 3–19.

Martin, John Jefferies. "Religion, Renewal and Reform in the Sixteenth Century." In *Early Modern Italy,* edited by John A. Marino, 166–87. Oxford: Oxford University Press, 2002.

Martin, Ruth. *Witchcraft and the Inquisition in Venice, 1550–1650.* Oxford: Basil Blackwell, 1989.

Matheson, Peter. *Cardinal Contarini at Regensburg*. Oxford: Clarendon, 1972.

Mayer, Thomas F. *Reginald Pole: Prince and Prophet*. Cambridge: Cambridge University Press, 2000.

McCluskey, Stephen. "Natural Knowledge in the Early Middle Ages." In *The Cambridge History of Science*. Vol. 2, *Medieval Science*, edited by David C. Lindberg and Michael H. Shank, 286–301. Cambridge: Cambridge University Press, 2013.

McConica, James. *Erasmus*. Oxford: Oxford University Press, 1991.

McKitterick, Rosamund. "The Church." In *The New Cambridge Medieval History*. Vol. 3, *c.900–c.1024*, edited by Timothy Reuter, 130–62. Cambridge: Cambridge University Press, 2000.

McMichael, Steven, and Susan Myers, eds. *Friars and Jews in the Middle Ages and Renaissance*. Leiden: Brill, 2004.

McNair, Philip. *Peter Martyr in Italy: An Anatomy of Apostasy*. Oxford: Oxford University Press, 1967.

Menchi, Silvana Seidel. *Erasmo in Italia (1520–1580)*. Turin: Bollati Boringhieri, 1987.

———. "Italy." In *The Reformation in National Context*, edited by Bob Scribner, Roy Porter, and Mikuláš Teich, 183–84. Cambridge: Cambridge University Press, 1994.

Menenghin, Vittorino. *I Monti di Pietà in Italia dal 1462 al 1562*. Vicenza: L.I.E.F., 1986.

Midelfort, H. C. Erik. "Witch Craze: Beyond the Legend of the Panic." *Magic, Ritual, and Witchcraft* 6 (2011): 11–33.

Minnich, Nelson H. *The Decrees of the Fifth Lateran Council (1512–17): Their Legitimacy, Origins, Contents and Implementation*. Abingdon: Routledge, 2016.

———. "Egidio da Viterbo, the Reform of Religious Orders, and the Fifth Lateran Council (1512–1517)." In *Egidio da Viterbo Cardinale Agostiniano tra Roma e l'Europa del Rinascimento: Atti del Convegno, Viterbo, 22–23 settembre 2012—Roma, 26–28 settembre 2012*, edited by Myriam Chiabò, Rocco Ronzani, and Angelo Maria Vitale. Rome: Centro Culturale Agostiniano and Roma nel Rinascimento, 2014.

———. *The Fifth Lateran Council (1512–17): Studies on Its Membership, Diplomacy and Proposals for Reform*. Aldershot: Ashgate Variorum, 1993.

Mixson, James D. "Religious Life and Observant Reform in the Fifteenth Century." *History Compass* 11 (2013): 201–14.

Mixson, James D., and Bert Roest, eds. *A Companion to Observant Reform in the Late Middle Ages and Beyond*. Leiden: Brill, 2015.

Monfasani, John. "Aristotelians, Platonists, and the Missing Ockhamists: Philosophical Liberty in Pre-Reformation Italy." *Renaissance Quarterly* 46 (1993): 247–76.

Monter, William. *Frontiers of Heresy: The Spanish Inquisition from the Basque Lands to Sicily*. Cambridge: Cambridge University Press, 1990.

Monter, William, and John Tedeschi. "Towards a Statistical Profile of the Italian Inquisitions: Sixteenth to Eighteenth Centuries." In *The Inquisition in Early Modern Europe: Studies in Sources and Methods*, edited by Gustav Henningsen and John Tedeschi, 130–57. DeKalb: Northern Illinois University Press, 1986.

Moore, Jill. *Inquisition and Its Organisation in Italy, 1250–1350*. Woodbridge: York Medieval, 2019.

Moore, R. I. *The Formation of a Persecuting Society: Authority and Deviance in Western Europe, 950–1250*. Paperback ed. Malden, MA: Blackwell, 1990.

———. *The War on Heresy: Faith and Power in Medieval Europe*. London: Profile Books, 2012.

Mormando, Franco. *The Preacher's Demons: Bernardino of Siena and the Social Underworld of Early Renaissance Italy*. Chicago: University of Chicago Press, 1999.

Morrall, John B. *Gerson and the Great Schism*. Manchester: Manchester University Press, 1960.

Morris, Colin. *Papal Monarchy: The Western Church from 1050–1250*. Oxford: Oxford University Press, 1989.

Muessig, Carolyn. "Bernardino da Siena and Preaching as a Vehicle for Religious Transformation." In Mixson and Roest, *Companion to Observant Reform*, 185–203.

Mulchahey, M. Michèle. *"First the Bow Is Bent in Study . . .": Dominican Education before 1350*. Toronto: Pontifical Institute of Medieval Studies, 1998.

Mullett, Michael A. *The Catholic Reformation*. London: Routledge, 1999.

Muraro, Luisa. *La Signora del Gioco: Episodi della caccia alle streghe*. Milan: Feltrinelli, 1976.

Murphy, J. C. "The Early Franciscan Studium at the University of Paris." In *Studium generale: Studies Offered to Astrik L. Gabriel*, edited by Leslie S. Domonkos and Robert J. Schneider, 159–203. Notre Dame, IN: Mediaeval Institute, University of Notre Dame, 1967.

Newman, William R. "Art, Nature and Demons: The Case of the *Malleus maleficarum* and Its Medieval Sources." In *The Artificial and the Natural: An Evolving Polarity*, edited by Bernadette Bensaude-Vincent and William R. Newman, 109–34. Cambridge, MA: MIT Press, 2007.

———. "Medieval Alchemy." In *Cambridge History of Science*. Vol. 2, *Medieval Science*, edited by David C. Lindberg and Michael H. Shank, 385–403. Cambridge: Cambridge University Press, 2013.

———. "An Overview of Roger Bacon's Alchemy." In Hackett, *Roger Bacon*, 317–36.

———. "The Philosopher's Egg: Theory and Practice in the Alchemy of Roger Bacon." *Micrologus* 3 (1995): 75–101.

———. "Technology and Alchemical Debate in the Late Middle Ages." *Isis* 80 (1989): 423–45.

Niccoli, Ottavia. *Prophecy and People in Renaissance Italy*. Translated by L. G. Cochrance. Princeton, NJ: Princeton University Press, 1990.

Noble, Thomas F. X. "The Papacy in the Eighth and Ninth Centuries." In *The New Cambridge Medieval History*. Vol. 2, *c.700–c.900*, edited by Rosamond McKitterick, 563–86. Cambridge: Cambridge University Press, 1995.

North, John. "Astronomy and Astrology." In *The Cambridge History of Science*. Vol. 2, *Medieval Science*, edited by David C. Lindberg and Michael H. Shank, 456–84. Cambridge: Cambridge University Press, 2013.

Nussdorfer, Laurie. "The Vacant See: Ritual and Protest in Early Modern Rome." *Sixteenth Century Journal* 18 (1987): 173–89.

Nutton, Vivian. "Medicine in Medieval Western Europe, 1000–1500." In *The Western Medical Tradition: 800 BC to AD 1800*, edited by Lawrence I. Conrad, Michael Neve, Vivian Nutton, Roy Porter, and Andrew Wear, 139–206. Cambridge: Cambridge University Press, 1995.

Oakley, Francis. "Conciliarism at the Fifth Lateran Council?" *Church History* 41 (1972): 452–63.

———. *The Conciliarist Tradition: Constitutionalism in the Catholic Church, 1300–1870*. Oxford: Oxford University Press, 2003.

O'Malley, John W. "The Feast of Thomas Aquinas in Renaissance Rome: A Neglected Document and Its Import." *Rivista di storia della chiesa in Italia* 35 (1981): 1–27.

———. *The First Jesuits*. Cambridge, MA: Harvard University Press, 1993.

———. *Giles of Viterbo on Church and Reform*. Leiden: Brill, 1968.

———. "Introduction." In O'Malley, *Collected Works of Erasmus*, 66:ix–li.

———. *Praise and Blame in Renaissance Rome: Rhetoric, Doctrine and Reform in the Sacred Orators of the Papal Court, c. 1450–1521*. Durham, NC: Duke University Press, 1979.

———. "Some Renaissance Panegyrics of Aquinas." *Renaissance Quarterly* 27 (1974): 174–92.

———. *Trent and All That: Renaming Catholicism in the Early Modern Era*. Cambridge, MA: Harvard University Press, 2000.

———. *Trent: What Happened at the Council*. Cambridge, MA: Belknap, 2013.

———. *What Happened at Vatican II*. Cambridge, MA: Belknap, 2008.

Page, Sophie. *Magic in the Cloister: Pious Motives, Illicit Interests, and Occult Approaches to the Medieval Universe*. University Park: Pennsylvania State University Press, 2013.

Page, Sophie, and Catherine Rider, eds. *The Routledge History of Medieval Magic*. London: Routledge, 2019.

Parente, Fausto. "The Index, the Holy Office, the Condemnation of the Talmud and Publication of Clement VIII's Index." In Fragnito, *Church, Censorship and Culture*, 163–93.

———. "La Chiesa e il 'Talmud': L'atteggiamento della Chiesa e del mondo cristiano nei confronti del Talmud e degli altri scritti rabbinici con particolare riguardo all'Italia tra XV e XVI secolo." In *Storia d'Italia*. Vol. 11, *Gli Ebrei in Italia, I, Dall alto Medio Evo all'età di Ghetti*, edited by C. Vivanti, 521–643. Turin: Einaudi, 1996.

Partington, J. R. "Albertus Magnus on Alchemy." *Ambix* 1 (1937): 3–20.

Paton, Bernadette. *Preaching Friars and the Civic Ethos: Sienna, 1380–1480*. London: Centre for Medieval Studies, Queen Mary and Westfield College, University of London, 1992.

———. "To the Fire, to the Fire! Let Us Burn a Little Incense to God: Bernardino, Preaching Friars and Maleficio in Later Medieval Siena." In *No Gods Except Me: Orthodoxy and Religious Practice in Europe, 1200–1600*, edited by Charles Zika, 8–36. Melbourne: History Department, University of Melbourne, 1991.

Pederson, Olaf. *The First Universities:* Studium Generale *and the Origins of University Education in Europe*. Cambridge: Cambridge University Press, 1997.

Pegg, Mark Gregory. *A Most Holy War: The Albigensian Crusade and the Battle for Christendom*. Oxford: Oxford University Press, 2008.

Peters, Edward. *Inquisition*. Berkeley: University of California Press, 1989.

———. *The Magician, the Witch, and the Law*. Hassocks: University of Pennsylvania Press, 1978.

Peterson, David S. "Religion and the Church." In *Italy in the Age of the Renaissance*, edited by John M. Najemy, 59–81. Oxford: Oxford University Press, 2004.

Pine, Martin. *Pietro Pomponazzi: Radical Philosopher of the Renaissance*. Padua: Antenore, 1986.

———. "Pomponazzi and the Problem of 'Double Truth.'" *Journal of the History of Ideas* 29 (1968): 163–76.

Porter, Martin. *Windows of the Soul: The Art of Physiognomy in European Culture, 1470–1780*. Oxford: Oxford University Press, 2005.

Principe, Lawrence M. *The Secrets of Alchemy*. Chicago: University of Chicago Press, 2013.

Prosperi, Adriano. "Italy." In *The Oxford Encyclopedia of the Renaissance*, vol. 2, edited by Hans J. Hillebrand, 324–29. Oxford: Oxford University Press, 1996.

———. *Tribunali della coscienza: Inquisitori, confessori, missionary*. New ed. Einaudi: Turin, 2009.

Prudlo, Donald S., ed. *The Origin, Development, and Refinement of Medieval Religious Mendicancies*. Leiden: Brill, 2011.

Pullan, Brian. *The Jews of Europe and the Inquisition of Venice, 1550–1670*. Oxford: Basil Blackwell, 1983.

Rawlings, Helen. *Church, Religion and Society in Early Modern Spain*. Houndsmills: Palgrave Macmillan, 2002.

———. *The Spanish Inquisition*. Malden, MA: Blackwell, 2006.

Resnick, Irvin M., ed. *A Companion to Albertus the Great: Theology, Philosophy and the Sciences*. Leiden: Brill, 2013.

Rider, Catherine. "Common Magic." In Collins, *Cambridge History of Magic*, 303–31.

Robinson, C. "Cosimo de Medici and the Franciscan Observants at Bosci ai Frati." In *Cosimo il Vecchio*, edited by Ames Lewes, 181–94. Oxford: Clarendon, 1992.

Robinson, I. S. "Reform and the Church, 1073–1122." In *The New Cambridge Medieval History*. Vol. 4, pt. 1, *c.1024–c.1198*, edited by David Luscombe and Jonathan Riley-Smith, 268–334. Cambridge: Cambridge University Press, 2004.

Robson, Michael. *The Franciscans in the Middle Ages*. Woodbridge: Boydell, 2006.

Rodriguez-Salgado, Maria. "The Habsburg-Valois Wars." In *The New Cambridge Modern History*. Vol. 2, *The Reformation, 1520–1559*, edited by Geoffrey R. Elton, 377–400. 2nd ed. Cambridge: Cambridge University Press, 1990.

Rollo-Koster, Joëlle. *Avignon and Its Papacy, 1309–1417: Popes, Institutions, and Society*. Lanham, MD: Rowman and Littlefield, 2016.

Ruggiero, Guido. *Binding Passions: Tales of Magic, Marriage and Power at the End of the Renaissance*. New York: Oxford University Press, 1993.

Rummel, Erika. *Jiménez de Cisneros: On the Threshold of Spain's Golden Age*. Tempe: Arizona Center for Medieval and Renaissance Studies, 1999.

Rusconi, Roberto. *L'ordine dei peccati: La confessione tra Medieoevo ed età*. Bologna: Il Mulino, 2002.

———. "Manuali milanesi di confessioni editi tra il 1474 e il 1523." *Archivium Franciscanum Historicum* 65 (1972): 107–56.

Russell, Camilla. *Giulia Gonzaga and the Religious Controversies of Sixteenth-Century Italy*. Turnhout: Brepols, 2006.

———. "Religious Reforming Currents in Sixteenth-Century Italy: The *Spirituali* and the Tridentine Debates over Church Reform." *Journal of Religious History* 38 (2014): 457–75.

Rutkin, H. Darrel. "Astrology." In *The Cambridge History of Science*. Vol. 3, *Early Modern Science*, edited by Katherine Park and Lorraine Daston, 541–61. Cambridge: Cambridge University Press 2006.

———. *Sapientia Astrologica: Astrology, Magic and Natural Knowledge, ca. 1250–1800*. Vol. 1, *Medieval Structures (1250–1500): Conceptual, Institutional, Socio-political, Theologico-religious, Cultural*. Cham: Springer, 2019.

Ryan, John J. *The Apostolic Conciliarism of Jean Gerson*. Atlanta, GA: Scholars, 1998.

Ryan, Michael A. *A Kingdom of Stargazers: Astrology and Authority in the Late Medieval Kingdom of Aragon*. Ithaca, NY: Cornell University Press, 2011.

Sackville, Lucy J. "The Church's Institutional Response to Heresy in the 13th Century." In *A Companion to Heresy Inquisitions*, edited by Donald S. Prudlo, 108–40. Leiden: Brill, 2019.

———. "The Inquisitor's Manual at Work." *Viator* 44 (2013): 201–16.

Schimmelpfennig, Bernhard. *The Papacy*. Translated by James Sievert. New York: Columbia University Press, 1992.

Seitz, Jonathan. *Witchcraft and Inquisition in Early Modern Venice*. Cambridge: Cambridge University Press, 2011.

Shapin, Steven. *Never Pure: Historical Studies of Science as If It Was Produced by People with Bodies, Situated in Time, Space, Culture and Society, and Struggling for Credibility and Authority*. Baltimore: Johns Hopkins University Press, 2010.

Siraisi, Nancy G. *Arts, Sciences at Padua: The Studium of Padua before 1350*. Toronto: Pontifical Institute of Medieval Studies, 1973.

———. *Medieval and Early Renaissance Medicine: An Introduction to Knowledge and Practice*. Chicago: Chicago University Press, 1990.

Stansbury, Ronald J. "Preaching and Pastoral Care in the Middle Ages." In *A Companion to Pastoral Care in the Middle Ages (1200–1500)*, edited by Ronald J. Stansbury, 23–39. Leiden: Brill, 2010.

Stow, Kenneth R. "The Burning of the Talmud in 1553, in the Light of Sixteenth-Century Catholic Attitudes toward the Talmud." *Bibliothèque d'Humanisme et Renaissance* 34 (1972): 435–59.

———. "The Church and the Jews." In *The New Cambridge Medieval History*. Vol. 5, *c.1198–c.1300*, edited by David Abulafia, 204–19. Cambridge: Cambridge University Press, 1999.

Sullivan, Karen. *The Inner Lives of Medieval Inquisitors*. Chicago: University of Chicago Press, 2011.

Swanson, R. N. "The 'Mendicant Problem' in the Later Middle Ages." In *The Medieval Church: Universities, Heresy and the Religious Life, Essays in Honour of Gordon Leff*, edited by Pete Biller and Barrie Dobson, 217–38. Woodbridge: Boydell, 1999.

Tarrant, Neil. "The Ambiguities of Censorship in Post-Tridentine Italy: The Case of Physiognomy." *Preternature* 8 (2019): 171–201.

———. "Between Aquinas and Eymerich: The Roman Inquisition's Use of Dominican Thought in the Censorship of Alchemy." *Ambix* 65 (2018): 210–31.

———. "Censoring Science in Sixteenth-Century Italy: Recent (and Not-So-Recent) Research." *History of Science* 52 (2014): 1–27.

———. "Giambattista Della Porta and the Roman Inquisition: Censorship and the Definition of Nature's Limits in Sixteenth-Century Italy." *British Journal for the History of Science* 46 (2013): 601–25.

———. "On the Origins of Enlightenment: The Fruits of Migration in the Italian Liberal Historiographical Tradition." In *Fruits of Migration: Heterodox Italian Migrants and Central European Culture, 1550–1620*, edited by Cornel Zwierlein and Vincenzo Lavenia, 362–81. Brill: Leiden, 2018.

———. "Reconstructing Thomist Astrology: Robert Bellarmine and the Papal Bull *Coeli et terrae*" *Annals of Science* 77 (2020): 26–49.

———. "Science, Religion and Italy's Seventeenth-Century Decline: From Francesco De Sanctis to Benedetto Croce." *Zygon* 54 (2019): 1125–44.

Tavuzzi, Michael. "Lorenzo Soleri da Sant'Agata OP (ob. ca. 1510): The 'Inquisitor Cumanus' of the *Malleus Maleficarum*—a Biographical Note." In Hoyer, *Praedicatores, Inquisitores I*, 407–20.

———. *Renaissance Inquisitors: Dominican Inquisitors and Inquisitorial Districts in Northern Italy, 1474–1527*. Leiden: Brill, 2007.

Tedeschi, John. "The Dispersed Archives of the Roman Inquisition." In *The Inquisition in Early Modern Europe: Studies in Sources and Methods*, edited by Gustav Henningsen and John Tedeschi, 13–32. DeKalb: Northern Illinois University Press, 1986.

———. "Inquisitorial Law and the Witch." In *Early Modern European Witchcraft: Centres and Peripheries*, edited by Bengt Ankarloo and Gustav Henningsen, 83–118. Oxford: Oxford University Press, 1990.

Théry, Julien. "A Heresy of State: Philip the Fair, the Trial of the 'Perfidious Templars' and the Pontificalization of the French Monarchy." *Journal of Medieval Religious Cultures* 39 (2013): 117–48.

Théry-Astruc, Julien. "The Pioneer of Royal Theocracy: Guillaume de Nogaret and the Conflicts between Philip the Fair and the Papacy." In *The Capetian Century, 1214–1314*, edited by William Chester Jordan and Jenna Rebecca Phillips, 219–59. Turnhout: Brepols, 2017.

Thijssen, J. M. M. H. *Censure and Heresy at the University of Paris, 1200–1400*. Philadelphia: University of Pennsylvania Press, 1998.

Thomson, Augustine. *Revival Preachers and Politics in Thirteenth-Century Italy*. Oxford: Clarendon, 1992.

Thorndike, Lynn. "Further Consideration of the *Experimenta, Speculum Astronomiae,* and *De Secretis Mulierum* Ascribed to Albertus Magnus." *Speculum* 30 (1955): 413–43.

Toaff, Ariel. "Jews, Franciscans and the First 'Monti di Pietà' in Italy (1462–1500)." In McMichael and Myers, *Friars and Jews*, 239–54.

Traver, Andrew. "The Forging of an Intellectual Defense of Mendicancy in the Medieval University." In Prudlo, *Origin, Development, and Refinement*, 157–96.

———. "Secular and Mendicant Masters of the Faculty of Theology at the University of Paris, 1505–1523." *Sixteenth Century Journal* 26 (1995): 137–55.

Tubau, Xavier. "Between Ecclesiology and Diplomacy: Francisco de Vargas and the Council of Trent." *Renaissance and Reformation* 42 (2019): 105–39.

———. "Hispanic Conciliarism and the Imperial Politics of Reform on the Eve of the Council of Trent." *Renaissance Quarterly* 70 (2017): 897–934.

———, ed. "Situating Conciliarism in Early Modern Spanish Thought." Special issue of *Renaissance and Reformation* 43 (2019): 9–166.

Utz Tremp, Kathrin. "Witches' Brooms and Magic Ointments: Twenty Years of Witchcraft Research at the University of Lausanne (1989–2009)." *Magic, Ritual, and Witchcraft* 5 (2010): 173–87.

Valliere, Paul. *Conciliarism: A History of Decision-Making in the Church*. Cambridge: Cambridge University Press, 2012.

Vanden Broecke, Steven. *The Limits of Influence: Pico, Louvain, and the Crisis of Renaissance Astrology*. Leiden: Brill, 2003.

Van Steenberghen, Fernand. *Aristotle in the West.* Translated by Leonard Johnson. Louvain: E. Nauwerlaerts, 1955.

———. *Thomas Aquinas and Radical Aristotelianism.* Washington, DC: Catholic University of America Press, 1980.

Vose, Robin. *Dominicans, Muslims and Jews in the Medieval Crown of Aragon.* Cambridge: Cambridge University Press, 2009.

———. "Heresy Inquisitions in the Later Middle Ages." In *A Companion to Heresy Inquisitions,* edited by Donald S. Prudlo, 141–71. Leiden: Brill, 2019.

Waley Singer, Dorothy. "Alchemical Writings Attributed to Roger Bacon." *Speculum* 7 (1932): 80–86.

Walsham, Alexandra. "The Reformation and the 'Disenchantment of the World' Reassessed." *Historical Journal* 51 (2008): 497–528.

Weill-Parot, Nicolas. *Les "images astrologiques" au Moyen Âge et à la Renaissance: Spéculations intellectuelles et pratiques magiques (XIIe–XVe siècle).* Paris: Champion, 2002.

Weinstein, Donald. *Savonarola: The Rise and Fall of a Renaissance Prophet.* New Haven, CT: Yale University Press, 2011.

Weisheipl, James A., ed. *Albertus Magnus and the Sciences: Commemorative Essays 1980.* Toronto: Pontifical Institute of Medieval Studies, 1980.

Westman, Robert S. *The Copernican Question: Prognostication, Skepticism, and Celestial Order.* Berkeley: University of California Press, 2011.

Wippel, John F. "The Condemnations of 1270 and 1277 at Paris." *Journal of Medieval and Renaissance Studies* 8 (1977): 169–201.

———. "Thomas Aquinas and the Condemnations of 1277." *Modern Schoolman* 72 (1995): 233–72.

Wright, Anthony D. *The Counter-Reformation: Catholic Europe and the Non-Christian World.* Frome: Butler and Tanner, 1982.

Yerushalmi, Hayim. "The Inquisition and the Jews of France in the Time of Bernard Gui." *Harvard Theological Review* 63 (1970): 317–76.

Zambelli, Paola. *The* Speculum astronomiae *and Its Enigma: Astrology, Theology and Science in Albertus Magnus and His Contemporaries.* Dordrecht: Kluwer, 1992.

Zutshi, P. N. R. "The Avignon Papacy." In *The New Cambridge Medieval History,* Vol. 6, *c.1300–c.1415,* edited by Michael Jones, 651–73. Cambridge: Cambridge University Press, 2000.

Index